The Scientific Bases for PRESERVATION of the MARIANA CROW

Committee on the Scientific Bases for the
Preservation of the Mariana Crow

Commission on Life Sciences

NATIONAL RESEARCH COUNCIL

NATIONAL ACADEMY PRESS
Washington, D.C. 1997

NATIONAL ACADEMY PRESS • 2101 Constitution Ave., N.W. • Washington, D.C. 20418

NOTICE: The project that is the subject of this report was approved by the Governing Board of the National Research Council, whose members are drawn from the councils of the National Academy of Sciences, the National Academy of Engineering, and the Institute of Medicine. The members of the committee responsible for the report were chosen for their special competences and with regard for appropriate balance. In preparing its report, the committee invited people with different perspectives to present their views. Such invitation does not imply endorsement of those views.

This report has been reviewed by a group other than the authors according to procedures approved by a Report Review Committee consisting of members of the National Academy of Sciences, the National Academy of Engineering, and the Institute of Medicine.

The National Academy of Sciences is a private, nonprofit, self-perpetuating society of distinguished scholars engaged in scientific and engineering research, dedicated to the furtherance of science and technology and to their use for the general welfare. Upon the authority of the charter granted to it by the Congress in 1863, the Academy has a mandate that requires it to advise the federal government on scientific and technical matters. Dr. Bruce Alberts is president of the National Academy of Sciences.

The National Academy of Engineering was established in 1964, under the charter of the National Academy of Sciences, as a parallel organization of outstanding engineers. It is autonomous in its administration and in the selection of its members, sharing with the National Academy of Sciences the responsibility for advising the federal government. The National Academy of Engineering also sponsors engineering programs aimed at meeting national needs, encourages education and research, and recognizes the superior achievements of engineers. Dr. Harold Liebowitz is president of the National Academy of Engineering.

The Institute of Medicine was established in 1970 by the National Academy of Sciences to secure the services of eminent members of appropriate professions in the examination of policy matters pertaining to the health of the public. The Institute acts under the responsibility given to the National Academy of Sciences by its congressional charter to be an adviser to the federal government and, upon its own initiative, to identify issues of medical care, research, and education. Dr. Kenneth I. Shine is president of the Institute of Medicine.

The National Research Council was organized by the National Academy of Sciences in 1916 to associate the broad community of science and technology with the Academy's purposes of furthering knowledge and advising the federal government. Functioning in accordance with general policies determined by the Academy, the Council has become the principal operating agency of both the National Academy of Sciences and the National Academy of Engineering in providing services to the government, the public, and the scientific and engineering communities. The Council is administered jointly by both Academies and the Institute of Medicine. Dr. Bruce Alberts and Dr. William A. Wulf are chairman and interim vice chairman, respectively, of the National Research Council.

This study by the National Research Council's Commission on Life Sciences was sponsored by the US Department of the Interior Fish and Wildlife Service under cooperative agreement no. 14-0001-48-96520. Points of view in this document are those of the authors and do not necessarily represent the official position of the US Department of the Interior.

Library of Congress Catalog Card Number 96-70149
International Standard Book Number 0-309-05581-4

Additional copies of this report are available from:
National Academy Press
2101 Constitution Ave., NW
Box 285
Washington, DC 20055
800-624-6242
202-334-3313 (in the Washington Metropolitan Area)
http://www.nap.edu

Copyright 1997 by the National Academy of Sciences. All rights reserved.

Printed in the United States of America

COMMITTEE ON THE SCIENTIFIC BASES FOR THE PRESERVATION OF THE MARIANA CROW

W. DONALD DUCKWORTH, Chair, Bishop Museum, Honolulu, Hawaii
STEVEN R. BEISSINGER, University of California, Berkeley
SCOTT R. DERRICKSON, National Zoological Park, Front Royal, Virginia
THOMAS H. FRITTS, National Biological Service, Washington, DC
SUSAN M. HAIG, Oregon State University, Corvallis
FRANCES C. JAMES, Florida State University, Tallahassee
JOHN M. MARZLUFF, Sustainable Ecosystems Institute, Meridian, Idaho
BRUCE A. RIDEOUT, Center for Reproduction of Endangered Species, San Diego, California

NRC Staff

ERIC A. FISCHER, Board Director
TANIA WILLIAMS, Study Director
PAULETTE A. ADAMS, Project Assistant
JEFFREY PECK, Project Assistant

BOARD ON BIOLOGY

MICHAEL T. CLEGG, *Chair*, University of California, Riverside, California
JOHN C. AVISE, University of Georgia, Athens
DAVID EISENBERG, University of California, Los Angeles
GERALD D. FISCHBACH, Harvard Medical School, Cambridge, Massachusetts
DAVID J. GALAS, Darwin Technologies, Seattle, Washington
DAVID V. GOEDDEL, Tularik, Inc., San Francisco, California
ARTURO GOMEZ-POMPA, University of California, Riverside
COREY S. GOODMAN, University of California, Berkeley
BRUCE R. LEVIN, Emory University, Atlanta, Georgia
OLGA F. LINARES, Smithsonian Tropical Research Institute, Miami, Florida
ELLIOTT M. MEYEROWITZ, California Institute of Technology, Pasadena
ROBERT T. PAINE, University of Washington, Seattle
RONALD R. SEDEROFF, North Carolina State University, Raleigh
DANIEL SIMBERLOFF, Florida State University, Tallahassee
ROBERT R. SOKAL, State University of New York, Stony Brook
SHIRLEY M. TILGHMAN, Princeton University, Princeton, New Jersey
RAYMOND L. WHITE, University of Utah, Salt Lake City

Staff:
ERIC FISCHER, Director
KATHLEEN BEIL, Administrative Assistant
JANET JOY, Program Officer
JEFFREY PECK, Project Assistant
TANIA WILLIAMS, Program Officer

COMMISSION ON LIFE SCIENCES

THOMAS D. POLLARD, *Chair*, Johns Hopkins University, Baltimore, Maryland
FREDERICK R. ANDERSON, Cadwalader, Wickersham & Taft, Washington, DC
JOHN C. BAILAR III, University of Chicago, Illinois
JOHN E. BURRIS, Marine Biological Laboratory, Woods Hole, Massachusetts
MICHAEL T. CLEGG, University of California, Riverside
GLENN A. CROSBY, Washington State University, Pullman
URSULA W. GOODENOUGH, Washington University, St. Louis, Missouri
SUSAN E. LEEMAN, Boston University, Massachusetts
RICHARD E. LENSKI, Michigan State University, East Lansing
THOMAS E. LOVEJOY, Smithsonian Institution, Washington, DC
DONALD R. MATTISON, University of Pittsburgh, Pennsylvania
JOSEPH E. MURRAY, Wellesley Hills, Massachusetts
EDWARD E. PENHOET, Chiron Corporation, Emeryville, California
EMIL A. PFITZER, Research Institute for Fragrance Materials Inc., Hackensack, New Jersey
MALCOLM C. PIKE, University of Southern California, Los Angeles
HENRY C. PITOT, III, University of Wisconsin, Madison
JONATHAN M. SAMET, Johns Hopkins University, Baltimore, Maryland
HAROLD M. SCHMECK JR, 32 Seapine Road, North Chatham, Massachusetts
CARLA J. SHATZ, University of California, Berkeley
JOHN L. VANDEBERG, Southwest Foundation for Biomedical Research, San Antonio, Texas

Staff:
PAUL GILMAN, Executive Director
SOLVEIG PADILLA, Administrative Assistant

Preface

The unique and fragile nature of environments found in oceanic island systems is well known. Through colonization and evolution, these relatively small land areas have evolved unique biota characterized by small populations, rare forms, and low phylogenetic diversity. They are also exceedingly vulnerable to human disturbance and to invasions of introduced species. Thus, while this report focuses on the challenges posed by preserving one species, the aga (or Mariana crow), it reflects the larger issues and challenges of biodiversity conservation in all oceanic island ecosystems.

The Mariana archipelago is the most northerly island group in Micronesia, lying roughly equidistant from Japan to the north, New Guinea to the south, and the Philippines to the west. Consisting of a north-south chain of 15 islands that extends over 675 km, the Marianas are in reality the emergent portions of a mighty mountain range that rises from the deepest portion of the Pacific Ocean, the Mariana Trench. The islands are volcanic in origin and generally diminish in size from south to north. The warm humid climate is distinguished by relatively constant temperature, wet and dry seasons, and a high annual probability of typhoons during the wet season.

The native ecosystems of the archipelago have been in decline since the earliest colonization by humans. Habitat destruction and alteration have been especially severe during this century as a result of war, post-war development, and associated intentional or accidental introduction of predatory or competitive species, especially on Guam. While these factors have impacted all groups of native organisms in the Mariana Islands to one degree or another, the extinction or near extinction of nearly all the native forest birds on Guam, most certainly linked to the large population of the introduced brown tree snake, is by far the most significant and alarming. The rapid decline of the aga is another painful reminder of the importance and challenge of preserving the unique environmental heritage of these islands, the need for increased knowledge to restore and maintain native species and habitats, and the compelling and lasting value of extensive public education to stimulate environmentally informed public policy development.

In response to a request from the US Fish and Wildlife Service (FWS), the National Research Council's Board on Biology established the Committee on the Scientific Bases for the Preservation of the Mariana Crow (aga) in March 1996. Its task was to review the existing data pertaining to the aga while focusing on several scientific issues:

- Assess to the extent possible the causes of the continuing decline in populations of the aga in the wild;
- Evaluate options for action to halt or reverse the decrease in numbers of the aga;
- Estimate the minimum viable population for survival of the Guam population of this species; and
- Evaluate the advisability of adding genetic material from the Rota population to the Guam population.

Current options for the recovery of the aga populations vary; they range from continuing present management strategies to a variety of translocations of one or both populations. From the assembled data and their evaluation, the committee has developed a set of recommendations designed to assist interested parties in working effectively to aid the recovery of the aga.

The committee held three meetings: one in Hawai'i (Honolulu, April 1996); one in Guam and Rota (May 1996); and one at the Arnold and Mabel Beckman Center in Irvine, California (June 1996). The committee collected and reviewed data on the aga from biologists in federal, state, territorial, and commonwealth agencies, and other scientists and individuals. Members visited the 'alala (Hawaiian crow) recovery project on the island of Hawai'i during the first meeting, and we are grateful to Cynthia Salley, McCandless Ranch, the FWS, and the Peregrine Fund for facilitating that opportunity. In Guam, the committee visited the laboratories and rearing facilities of both federal and Guam agencies and was extensively briefed both in field locations and formal settings. On Rota, nesting sites were observed and a broad range of aga habitat was visited through the support and generosity of Stan Taisacan of the Commonwealth of the Northern Mariana Islands Department of Lands and Natural Resources Division of Fish and Wildlife, and Dan Grout of FWS. The information and support provided by those people and many others were invaluable to the committee and to the development of this report. In addition, we are grateful to the many other persons and organizations who assisted the committee in preparing this report (See appendix B).

From the outset of this study, the committee has been constantly motivated to produce a report in a time-frame that would allow for consideration of its recommendations by the FWS and related agencies for implementation during this year's breeding season. Consequently, the committee schedule was fast-paced and intensive, but never at the expense of limiting the fullest measure of discussion and evaluation of available data in reaching consensus. It was my pleasure and privilege to work with a talented, enthusiastic, and dedicated committee. This report is the product of their hard work and spirited, yet always collegial, debate. I want to express my deepest gratitude for their patience, their insight, and their generosity in response to the rigorous schedule and scientific challenges imposed by the task.

Last, on behalf of the entire committee, I want to express our gratitude to the National Research Council staff for the vital role they played in all aspects of the study. The dedication and persistence of project director Tania Williams facilitated and expedited the committee's activities throughout the study. In addition, Paulette Adams and Jeff Peck provided support for our various activities.

W. Donald Duckworth, *Chair*
Committee on the Scientific Bases for
the Preservation of the Mariana Crow
September 1996

Contents

EXECUTIVE SUMMARY   1

1   INTRODUCTION   5
    Distribution and Abundance, 5
    Legal Status, 6
    Genetics, 8
    This Study, 8

2   HISTORY   11
    Description of Guam and Rota, 11
    Ecological History and Faunal Change, 15
    Natural History of the Aga, 19
    Comparison with the 'Alala, 21
    Demography of the Aga, 21
    Possible Causes of Population Declines, 26
    Population Projection for the Aga on Guam, 31
    Past and Current Conservation and Management, 31

| 3 | BROWN TREE SNAKE | 39 |

Natural History, 39
Effects of the Presence of the Brown Tree Snake on Guam, 42
Decline of Birds on Guam in Relation to the Brown
  Tree Snake, 45
Efforts to Control the Brown Tree Snake on Guam, 46
Prognosis for Control on Guam, 50
Threat of Spread of the Brown Tree Snake to Other
  Islands, 50

| 4 | MAJOR FINDINGS AND CRITERIA FOR RECOVERY OF THE AGA | 53 |

Summary of Major Findings, 53
Criteria for Prioritizing Aga Recovery Actions, 57

| 5 | OPTIONS FOR THE MANAGEMENT OF THE AGA | 59 |

Cessation of Management, 59
Continuation of Present Management Strategy, 59
Intensive Management, 60
Translocation Options, 62

| 6 | ELEVEN RECOMMENDED MANAGEMENT ACTIONS FOR RECOVERY AND THEIR RATIONALE | 67 |

Recommendation 1, 67
Recommendation 2, 69
Recommendation 3, 70
Recommendation 4, 70
Recommendation 5, 71
Recommendation 6, 71
Recommendation 7, 72
Recommendation 8, 72
Recommendation 9, 72
Recommendation 10, 73
Recommendation 11, 73

REFERENCES — 75

APPENDIX A: STATEMENT OF TASK — 87

APPENDIX B: ACKNOWLEDGMENTS — 88

APPENDIX C: COMMITTEE AND STAFF BIOGRAPHICAL INFORMATION — 90

Executive Summary

The aga, or Mariana crow (*Corvus kubaryi*), is a tropical forest crow that occurs only on the islands of Guam and adjacent Rota. Its precipitous decline on Guam in the last 20 years has been coincident with the extinction or near extinction of all the other native forest birds and with the population explosion of the introduced brown tree snake (*Boiga irregularis*). Electric barriers on the trunks of trees in which crows are attempting to nest can lower predation on eggs and nestlings by brown tree snakes and monitor lizards (*Varanus indicus*). However, even with such protection, a high percentage of eggs fail to result in fledglings. With only about 20 adult aga remaining in forested areas of northern Guam, and no remaining pairs known to lay viable eggs, it is doubtful that this population can be saved from extinction without the addition of birds from elsewhere. Several hundred aga are present on Rota, but that aga population has declined by at least 50% since the early 1980s. The brown tree snake is not known to be present.

The Guam Department of Agriculture Division of Aquatic and Wildlife Resources (DAWR), the Commonwealth of the Northern Mariana Islands (CNMI) Department of Lands and Natural Resources Division of Fish and Wildlife (DFW), and the US Department of the Interior Fish and Wildlife Service (FWS) and National Biological Service (NBS) are all committed to the preservation of the native fauna of the entire Mariana archipelago. Recognizing the need for a coordinated recovery program for the aga, FWS requested that a committee be appointed by the National Research Council to make appropriate scientific recommendations for its preservation. Thanks to the cooperation of the aforementioned agencies, the committee was able to review and analyze all available scientific information, including published papers, unpublished manuscripts, and internal reports. It was briefed by biologists from DAWR, DFW, FWS, and NBS, and spoke with other persons with relevant expertise, especially US Department of Agriculture Animal Damage Control (ADC) staff on Guam.

The committee concurs with the conclusion of Savidge and subsequent workers that the major cause of the aga's decline on Guam is predation by the brown tree snake. The probable presence of introduced populations of the brown tree snake on other islands within the Mariana archipelago and elsewhere in the Pacific portends a sequence of severe ecological disruptions; this is a matter potentially far more serious than the preservation of the aga on Guam and Rota. Development and implementation of effective, large-scale methods for controlling the brown tree snake are urgently needed to prevent the further spread of this Australasian species to other islands and to enable the recovery of the aga population on Guam. At present, the ADC program constitutes the only impediment to dispersal of the snake from Guam.

The cause of the high percentage of nonviable eggs of the aga on Guam is less clear. The cause of the decline of the aga on Rota is also poorly understood, although the reduction of forested habitats and the direct killing of crows have obviously been contributing factors. On both islands, a segment of the public considers economic development to have a generally higher priority than preservation of habitat for endangered and threatened species. Public leaders on Rota, however, have demonstrated increasing concern for environmental quality and preservation. Several protected areas have been established recently, and additional areas have been proposed for protection. CNMI, in cooperation with FWS personnel, is preparing a habitat-conservation plan for Rota that will designate areas for conservation and for potential development; it is hoped that this plan will serve as a model for conservation planning on other islands in the Pacific basin. Clearly, a viable aga population on Rota, Guam, or elsewhere cannot be maintained unless adequate habitat is preserved.

There are 12 aga in captivity. Between 1993 and 1995, ten were collected from Rota, of which nine are still alive. Additionally, one young bird, which was produced by a pair housed at the Houston Zoological Gardens in Texas, remains at that facility. Two young hatched in 1996 from eggs taken from nests of wild aga on Guam are now being held in the DAWR facility there. Because substantial numbers of birds still exist on Rota, the committee believes that an intensive captive-breeding program, like the one for the 'alala in Hawaii, is not justified at this time. Similarly, although it is important to maintain two wild populations, the committee believes that management to preserve the apparent low level of genetic differences between the Guam and Rota populations is not justified. Instead, emphasis should be on implementing a coordinated program of research and management that will alleviate the causes of declines in both wild populations, stabilize the Rota population at a size and distribution that are viable in the longterm, and sustain the Guam population and allow it to increase as quickly as possible. However, because the Rota population is the largest and most-secure aga population, the committee recommends that the highest priority be given to research and management actions necessary for the long-term preservation of the Rota population while attempts to preserve and restore the Guam population proceed.

In attempting to formulate such a coordinated recovery program for these aga populations, the committee has considered various options, including implementation of large-scale snake-control measures in native habitat, intensive avicultural management, interisland translocations, and return of captive crows to the wild. After considering those options and acknowledging the urgent need for both short- and long-term actions, the committee formulated the following eleven recommendations for steps to be taken over the next three years. Chapter 6 presents a complete exposition of these recommendations.

Recommendation 1. Expand and increase research, development, and implementation of methods to control the brown tree snake and to prevent its spread to other islands.

Recommendation 2. Study the behavior, population biology, and health of marked birds on both Guam and Rota.

Recommendation 3. Release all twelve captive aga on Guam within the next two years.

Recommendation 4. Place electric barriers in all nest trees to protect nests from brown tree snakes and monitor lizards, and increase trapping efforts in nest trees and adjacent areas.

Recommendation 5. Assess the feasibility of translocating birds to islands other than Guam or Rota.

Recommendation 6. Conduct annual censuses of the aga population on both islands.

Recommendation 7. Conduct complete postmortem examinations on all recoverable eggs and birds.

Recommendation 8. Facilitate the development of a habitat conservation plan for Rota.

Recommendation 9. Establish a professional-level research manager position within the National Biological Service on Guam.

Recommendation 10. Appoint a recovery team specifically for the aga.

Recommendation 11. Conduct a public education program that specifically addresses issues relevant to aga conservation, including the problems associated with the spread of the brown tree snake.

1

Introduction

The Mariana crow (*Corvus kubaryi*), or aga in the Chamorro language, is endemic to the two southernmost islands in the Mariana archipelago, Guam and Rota, and is a federally listed endangered species. The current rate of decline in the Guam population of the aga is serious: in the opinion of the US Fish and Wildlife Service (FWS), the population will soon become extinct unless appropriate action is taken. The Rota population, larger than that of Guam, has probably also declined in recent years. Forested crow habitat is being converted to farmland and developed for other uses associated with the island's increasing tourism and population growth. Guam and Rota, separated by 49 km of ocean, have distinct but related government jurisdictions. Guam is administered as a US territory, and Rota and the other islands in the archipelago form the Commonwealth of the Northern Mariana Islands (CNMI). In any case, the native people of the archipelago, the Chamorros, share a common bond of language and culture throughout the islands.

DISTRIBUTION AND ABUNDANCE

The aga has been restricted to Guam and Rota in historical times, although there is some evidence that it is a relict species that formerly had a wider distribution (Baker 1951). Formerly, the aga was distributed throughout forested areas on both islands. At one point, crows on Guam were considered abundant enough to be an agricultural pest (Safford 1902; Seale 1901). Likewise, on Rota, they were common enough to be hunted by visitors from Saipan even into the 1970s (Pratt and others 1979). In 1945, after World War II, Baker (1951) found the crows to be common in forested areas and coconut plantations on Guam but uncommon in areas densely populated by humans. Crows disappeared from southern Guam in the middle 1960s and from central Guam in the middle 1970s (FWS 1984), and now they remain only in the northern part of the island on Andersen Air Force Base. No aga chicks have successfully fledged from unprotected nests since 1987, and only six have fledged from protected nests since 1989. The single chick that fledged in 1989 from a nest tree protected with a steel-wrap and tanglefoot barrier was killed soon after fledging by a monitor lizard. Five other chicks were fledged from nest trees protected with electric barriers between 1992 and 1994 (Aguon and others in press; Wiles and others 1995). Two additional chicks, collected from wild nests in the 1995–1996 breeding season, were successfully hatched and hand-raised in captivity and are being held at the Guam Department of Agriculture Division of Aquatic and Wildlife Resources (DAWR) facilities (Beck and Brock 1996). The aga population on Guam has declined rapidly from about 100 birds in the mid-1980s to about 25 birds at present (Table 1-1). Interestingly, the Rota population appears to have declined substantially in recent years as well (Table 1-1).

The wild aga population on Guam numbers about 20 (DAWR 1996), and observational evidence gathered by DAWR personnel suggests a sex ratio skewed in favor of males. The population is restricted to the extreme northern portions of Guam, an area encompassing less than one-fourth of the crow's original distribution on the island. On Rota, crows still occur in suitable habitat throughout the island, although the population has clearly declined since an islandwide survey was completed in 1982 (Engbring and others 1986; Grout and others 1996). A survey conducted by FWS in October–November 1995 yielded an estimated population of 592 (SE = 63, 95% confidence interval = 474–720) (Grout and others 1996). A reanalysis of the 1982 survey data suggests a 56% decline in the Rota population from 1982 to 1995 (Grout and others 1996).

Although available survey data suggest a general decline in both wild populations (Table 1-1), these data are difficult to assess because census techniques and census efforts have varied. The accuracy of the variable-circular-plot (VCP) method can also be questioned for crows because their calls can be heard from considerable distances (Engbring and Ramsey 1984). The discrepancies between the number of birds observed and the number of birds estimated suggests that current population estimates are not as accurate as previously reported.

LEGAL STATUS

The first legal protection of the aga occurred on Guam in 1981 with the passage of Guam Public Law 16-39. It was declared an endangered species by Guam (Endangered Species Act of Guam, Guam Public Law 15-36) in June 1981 and later was given endangered status on Guam and Rota under the US Endangered Species Act in 1984 with six other avian and two mammalian species

TABLE 1-1 Estimated Populations of Aga on Guam and Rota

	Guam		Rota	
Year	Estimate	Source	Estimate	Source
1901	Abundant in forests	Seale 1901	—	—
1945	Abundant in forests	Baker 1951	—	—
1981	357	Engbring and Ramsey 1981	—	—
1982	—	—	1,318 (454)[a]	Engbring and others 1986
1985	100	Michael 1987	—	—
1988	—	—	600–1,000	R. Beck and S. Pimm (unpublished data) in Aguon and others (in press)
1990–1991	107	DAWR summary	—	—
1991	41	DAWR summary	2,132 (67)[a]	M. Lusk 1995 Sep 22[b]
1992	57	DAWR summary	447–931 (26–84)[a]	M. Lusk 1995 Sep 22[b]
1993	51	DAWR summary	800	CNMI DFW annual report fiscal year 1995
1994	40–50	DAWR summary	336–454 (31–65)[a]	M. Lusk 1995 Sep 22[b]
1995	26	DAWR summary	592	Grout and others 1996

[a]Observed.
[b]Draft memo on analysis of Rota crow population survey from M. Lusk to K. Rosa, Chief Recovery Branch, US FWS Pacific Region.

(50 CFR 17; 49 FR 33881). Critical habitat on Guam was not designated under the original federal listing but was proposed for two species of bats and four species of birds, including the aga, in June 1991 (FWS 1991). However, FWS withdrew the proposal in 1994 (FWS 1994) after the establishment of the Guam National Wildlife Refuge. The refuge encompasses an overlay of 9,100 ha of cooperatively managed lands owned by FWS, the US Navy, and the US Air Force. The aga is also listed as endangered by CNMI.

GENETICS

An understanding of the genetic relationship between the Guam and Rota populations is critical to evaluating conservation options. Tarr and Fleischer's (1995) mitochondrial DNA (mtDNA) and DNA "fingerprinting" work on aga, American crows (*Corvus brachyrhynchos*), and ravens (*Corvus corax*), although limited, sheds light on this issue. The mtDNA data reveal only one polymorphic site for aga, and each insular population was fixed for all alternate bases at that site. In addition, the total sequence difference between populations was only about one-tenth that between raven populations. Samples were small (8 Guam birds and 10 Rota birds), but DNA fingerprinting indicated close similarity between the populations. Thus, although the two populations are genetically distinct, the differences are very small, and the colonization of Rota from Guam might have been relatively recent.

The genetic diversity within the two populations is of additional interest. The Guam sample included close relatives, but the extent of DNA band-sharing among them was much lower (that is, more genetically diverse: $S = 40$–47%) than that among the Rota birds, which were sampled from various locations over the island ($S = 69$–88%). The greater genetic diversity of the Guam population suggests that the Rota population was derived from a fairly recent (on an evolutionary scale) colonization of Rota by relatively few individuals from Guam and that there has been relatively little interchange between populations. Management implications of those findings will be discussed in chapter 5.

THIS STUDY

Responsibility for the recovery of the aga under the Endangered Species Act rests with FWS. FWS asked the National Research Council to conduct a scientific review of the information available on the aga and to recommend appropriate specific recovery actions to prevent extinction of the Guam population and to ensure species survival. The Committee on the Scientific Bases for Preservation of the Mariana Crow—comprising experts in ornithology, conservation biology, captive propagation, herpetology, avian pathology, behavior, and population ecology—was formed under the Research Council's Board on Biology and charged in March 1996 to review the available relevant environmental information and to prepare a report detailing the status of the two populations of the aga and the management actions necessary for their recovery and conservation (see appendix A, "Statement of Task"). The committee examined the relevant scientific data, held a public meeting, and spoke to numerous persons, including biologists with federal agencies, DAWR, and the CNMI Department of Lands and Natural Resources' Division of Fish and Wildlife (see appendix B, "Acknowledgments"). The committee also visited study sites and research facilities on both Guam and Rota.

This report reflects the committee's analysis of all available information on the aga. Chapter 2 summarizes what is known and identifies what is unknown about the aga's history and decline; it includes a comparison with the 'alala, a demographic analysis, a discussion of extrinsic factors that might have contributed to the decline of the aga, and past conservation and management efforts in behalf of both wild and captive crows. Chapter 3 provides an extensive review of the brown tree snake (*Boiga irregularis*), the cause of the aga's decline on Guam and a substantial threat to the aga's long-term survival on Rota. Chapter 4 summarizes the committee's major findings and criteria for recovery of the aga. Chapter 5 describes options for management of the aga. Finally, Chapter 6 recommends eleven specific management actions that the committee believes are essential for the aga's recovery.

2

History

The aga occurs only on 2 islands in the Mariana archipelago: Guam and Rota. Although these islands are similar in their origins and overall ecology, they differ considerably in some of their physical characteristics and in their ecological and political histories. To place the current status of the Guam and Rota populations of the aga within a historical framework, the initial sections of this chapter describe the physical and climatic characteristics of both islands, provide an overview of the ecological changes that have occurred from human colonization to the present, and summarize recent avian population trends and extinctions on each island. Later sections focus specifically on the aga, providing an overview of its natural history, probable causes of decline, population demography, and past and present conservation measures.

DESCRIPTION OF GUAM AND ROTA

Micronesia encompasses an area of 7.5 million square kilometers in the western Pacific that stretches from the equator to 20° N and from the International Date Line to 130° E. Although more than 2,000 islands are in Micronesia in the Gilbert, Marshall, Caroline, and Mariana islands, the total combined land area of all the islands is less than 2,000 km^2 (Engbring and Pratt 1985; Engbring and others 1986), which is smaller than the state of Rhode Island (3,144 km^2). The Mariana archipelago is the most northerly island group in Micronesia, lying midway between Japan and New Guinea and about 2,600 km east of the Philippines. Volcanic in origin, the 15-island archipelago has a total land area of about 640 km^2 (Baker 1951) and extends over 675 km from Guam (13° 30' N, 144° 45' E) in the south to Uracas (20° 31' N, 144° 54' E) in the north (Figure 2-1). The islands generally decrease in size from south to north, and some of the northern islands are actively volcanic (Reichel and Glass 1991). Guam is the most southerly and populous of the Mariana Islands and is administered as a US territory. All the other islands in the archipelago are part of the Commonwealth of the Northern

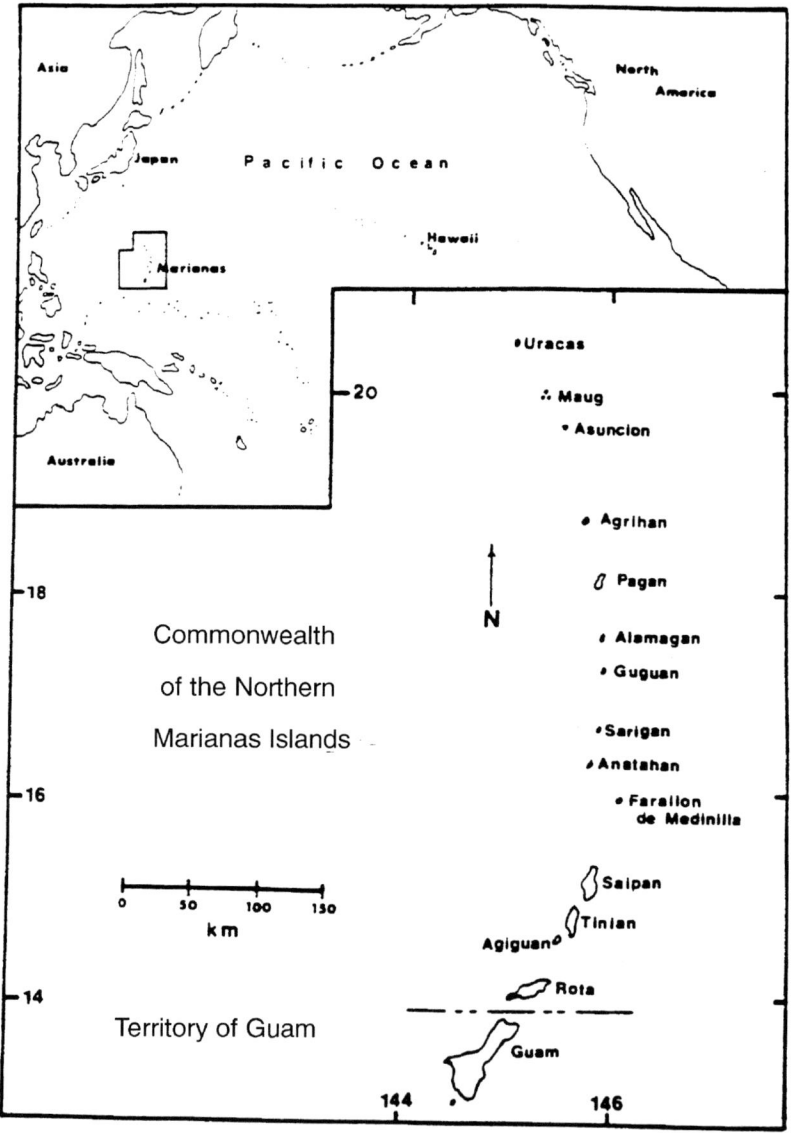

FIGURE 2-1 Mariana Archipelago. Source: Adapted from Engbring and others 1986.

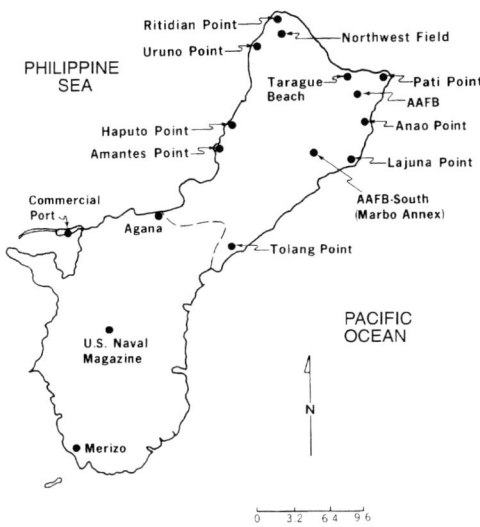

FIGURE 2-2 Guam. Dashed line indicates the approximate area of transition from the northern plateau to the southern mountainous area. Source: Adapted from Jenkins 1983.

Mariana Islands (CNMI). Saipan is the most populous island of the CNMI, followed by Rota and Tinian. Agiguan and the ten islands north of Saipan are small and now uninhabited (Engbring and Pratt 1985; Engbring and others 1986).

Guam (Figure 2-2) is the largest island (550 km^2) in the Mariana archipelago, and the most developed island in Micronesia. About 45 km long and 6–13 km wide, it is separated into a northern, uplifted limestone plateau and a southern mountainous region of volcanic origin (FWS Figure 2-1, Figure 2-2 1990). The northern plateau averages 100–200 m in elevation, whereas the highest mountain (Mount Lamlam) in the south is 405 m. The island is largely surrounded by fringing reefs except for small areas along the steep limestone cliffs of the northern plateau. Because of the porous coralline limestone soil, the northern plateau lacks permanent streams and marshes. In contrast, streams, ponds, and marshes are common in the south, where the volcanic soil is nonporous.

Guam is believed to have been covered by tropical broadleaf forests before human colonization (Fosberg 1960). Today, however, the island is greatly dissected by fields, savannas, urban and suburban areas, commercial facilities, military installations, and roads. Although the northern plateau still supports relatively diverse secondary "limestone" and scrub forest, primary forest is largely restricted to the cliffline and immediately adjacent areas (Conry 1988; Engbring and Ramsey 1984; Jenkins 1983; Stone 1970). In the south, the uplands are dominated by savannas and the sheltered valleys by ravine forest (Stone 1970). The extensive savannas probably resulted from clearing by the island's first inhabitants and a burning regime (Fosberg 1960; Mueller-Dumbois 1981).

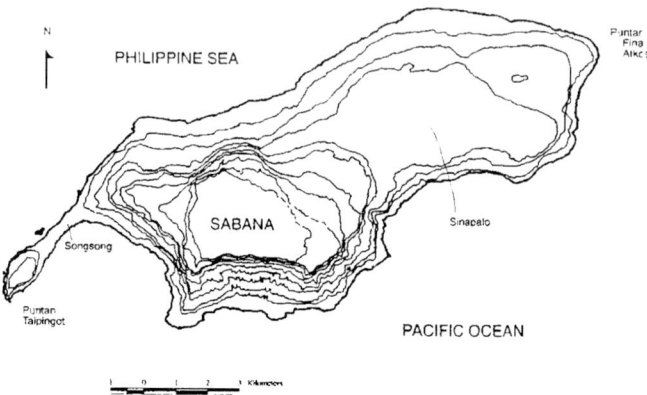

FIGURE 2-3 Island of Rota, Commonwealth of the Northern Mariana Islands. The majority of the island's inhabitants reside in the villages of Songsong and Sinapalo. Source: Committee-generated from USGS topographical maps.

Rota lies about 49 km north of Guam (Figure 2-1). It is about 20 km long and 4–8 km wide, with an area of about 85 km^2 (FWS 1990). Much of the island's coastline is rocky and steep, although there is a substantial coral sand beach inside the fringing reef along most of the northern shoreline. The village of Songsong and two commercial ports are on the isthmus of a prominent, narrow peninsula that juts to the southwest from the western side of the island (Figure 2-3). The topography of the western half of the island is dominated by a large uplifted plateau (12 km^2) known as the Sabana at an elevation of about 450 m. The plateau is mostly open grassland as a result of past agricultural and phosphate-mining activities, and it is still used extensively for agricultural production. Steep cliffs drop from the north, west, and south sides of the Sabana to narrow coastal shelves and then to the ocean. The northeastern side of the Sabana slopes more gently to a large shelf with an elevation of about 150 m, which covers the eastern half of the island. This area and adjacent coastal shelves were cultivated for sugar cane during the Japanese administration of the island, in 1914–1944 (Engbring and others 1986). Although many of those areas have reverted to scrubby second-growth forest since World War II, they are now being progressively cleared for agricultural homesteads, grazing, and tourist development. The native vegetation remains less disturbed on Rota than on Saipan or Tinian because much of Rota's terrain is precipitous and unsuitable for large-scale agriculture and because it suffered less damage during World War II. Vegetation and habitat types on Rota are similar to those on Guam (Engbring and others 1986; Fosberg 1960).

The Mariana Islands are warm and humid with only moderate seasonal or daily variation in temperature. Temperatures usually range from 25 to 32°C during the day and from 24 to 27°C at night. The mean annual temperature is about 27°C on Guam and slightly lower on Rota (Eldredge 1983). Annual rainfall is about 220 cm; three-fourths of the total falls during the wet season (July–November) (Engbring and Pratt 1985; Engbring and Ramsey 1984; Engbring and others 1986; Stone 1970). Both temperature and rainfall decrease from south to north within the Mariana archipelago (Butler 1991). Winds can blow from any direction during the wet season but are generally light. The dry season (January–May), which can vary considerably in length and intensity, is characterized by steady easterly and northeasterly trade winds (Engbring and Ramsey 1984). Major annual variations in weather patterns are related primarily to the distribution of rainfall and the frequency of typhoons. Typhoons are common in this region of the Pacific during the wet season, and there is a 1 in 3 chance that a typhoon will pass close enough to affect the inhabited islands of the Marianas in any particular year (NOAA 1982).

ECOLOGICAL HISTORY AND FAUNAL CHANGE

As in other isolated and small island groups of the Pacific Basin, the prehuman fauna of the Mariana Islands was shaped primarily by infrequent immigration and colonization events (Diamond and Mayr 1976; MacArthur and Wilson 1967). Most of the endemic land birds appear to have been derived from colonizations from the Philippine and New Guinea regions (Baker 1948, 1951). Endemic mammals included only fruit bats (*Pteropus mariannus* and *P. tokudae*) and the sheath-tailed bat (*Emballonura semicaudata*). Knowledge of the historical reptilian and invertebrate fauna is poor, but it seems likely that the endemic fauna was restricted to the blind snake (*Rantophotylops brahminus*), several species of skinks and geckos, and a number of partulid snails, and numerous insects (Rodda and Fritts 1992; Wiles and others 1990, 1995).

The Chamorros colonized the Mariana Islands from the Indo-Malaysian region about 4,000 years ago (Craib 1983). A variety of plant and animal commensals were introduced at that time, including such food plants as rice (Solenberger 1967) and such animals as the red jungle fowl (*Gallus gallus*) and rats (*Rattus exulans* and possibly *R. rattus*). The southern "high islands" of Guam, Rota, Agiguan, Tinian, and Saipan are believed to have been completely forested at the time of colonization (Fosberg 1960). As the human population increased, slash-and-burn methods for cultivation of food crops led progressively to the development of open savannas in upland areas (Mueller-Dumbois 1981). Recent paleontologic excavations on Rota, Agiguan, and Tinian have revealed that, as elsewhere on oceanic islands, numerous species of land birds and resident breeding populations of sea birds were extirpated in prehistoric times as a result of overharvesting by humans and predation by introduced predators (Atkinson 1984; Steadman 1995a). Excavations on Rota have yielded bones of 22 resident avian species in the late Holocene (last 5,000 years), of which 13 are no longer on the island

(Steadman 1992, 1995b). It is expected that the number of extinct and extirpated species will increase as additional data are obtained.

Magellan's "discovery" of the Mariana Islands in 1521 ushered in an era of ecological change that continues today. In 1521, at least 13 islands were inhabited by the native Chamorro people, and their population is estimated to have been between 30,000 and 40,000. By the early 1700s, the Chamorro population had been reduced to less than 4,000 by epidemics, war, and their forced relocation by the Spanish from northern islands to Guam and other Spanish colonies (Carano and Sanchez 1964; Hezel and Berg 1981). Although the Spanish presence in the Marianas was sporadic until a permanent mission was established on Guam in 1668, the Marianas were visited regularly for water and provisions by galleons traveling between the Americas and the Philippines after 1565 (Hezel 1982). A number of plants and animals were introduced to the islands from the Philippines and Mexico (Butler 1991; Engbring and others 1986; Hezel 1982; Seale 1901) as a result of the galleon trade (for example, maize, *Zea*; goat, *Capra* sp.; cattle, *Bos* sp.; pig, *Sus scrofa*; and deer, *Cervus unicolor*) and later increased commerce with the Philippines (for example, Philippine turtle dove, *Streptopelia bitorquata*; blue-breasted quail, *Coturnix chinensis*; water buffalo, *Bubalus bubalis*; Norway rat, *Rattus norvegicus*; house mouse, *Mus musculus*; monitor lizard, *Varanus indicus;* domestic cat, *Felis cattus;* and domestic dog, *Canis familiaris*). The introduced ungulates had an especially dramatic effect on the native forest and understory, and introduced predators undoubtedly had an adverse effect on the native vertebrate fauna.

After the Spanish-American War (1898), Spain ceded Guam to the United States; except for a brief period of Japanese occupation in World War II (1942–1944), Guam has remained a US territory. In 1899, Spain sold all the other Mariana Islands to Germany. The period of German administration of the northern islands, 1899–1914, saw relatively few ecological changes, although there was a substantial immigration from the Carolines to some islands north of Guam and an increase in the development of coconut plantations and orchards (Engbring and others 1986).

After World War I, Germany relinquished its control of the northern islands to Japan under a mandate of the League of Nations. The period of Japanese administration of the northern islands, 1914–1944, witnessed extensive selective logging, phosphate mining, and agricultural development and an associated increase in immigration of Japanese, Korean, and Okinawan workers. On Saipan, Tinian, Agiguan, and Rota, native forest was cleared on nearly all flat, tillable areas for sugar-cane production (Baker 1946; Hezel and Berg 1981; Mayr 1945). For example, on Rota, a railroad was built that ringed the island, a sugar-cane mill was constructed in Songsong, and extensive phosphate mining was undertaken on the Sabana. Several conspicuous introductions of plants and animals occurred during this period (Baker 1951; Engbring and others 1986), including the ironwood (*Casuarina equisetifolia*), Formosan koa (*Acacia confusa*), flame tree (*Delonix regia*), black drongo (*Dicrurus macrocercus*), marine toad (*Bufo marinus*), and African land snail (*Achatina fulica*). The drongo, introduced to Rota from Formosa in 1935, colonized Guam in 1960 and is now well established on both islands (Jenkins 1983).

When US troops invaded the Marianas during World War II, the fighting, military construction, and influx of large numbers of men and material had many ecological effects on the southern islands (Baker 1946; Mayr 1945). On Saipan and Tinian, much of the remaining forest was damaged by the fighting. On Guam, habitat destruction associated with the fighting was less because organized Japanese resistance did not last as long. However, as on Tinian, large portions of Guam's native limestone forest were cleared for military construction (Fosberg 1960). Neither Rota nor Agiguan was invaded, but large areas of both islands were heavily bombed before Japanese surrender (Baker 1946; Butler 1991). The military seeded large areas with tangantangan to prevent erosion after invasion, and large areas are now overgrown with this tree and an assorted mixture of native and exotic vegetation (Engbring and others 1986). Since the war, a number of additional species have been either accidentally or intentionally introduced to Guam (Engbring and others 1986), including the brown tree snake (*Boiga irregularis*), musk shrew (*Suncus murinus*), black francolin (*Francolinus francolinus*), Eurasian tree sparrow (*Passer montanus*), and chestnut mannikin (*Lonchura molacca*). The brown tree snake apparently arrived from Melanesia in the late 1940s or early 1950s in war material that was being consolidated on Guam by the military (Rodda and others 1992). The brown tree snake, tree sparrow, and musk shrew have since spread to other islands in the archipelago.

Economic development on Guam has continued steadily since World War II as a result of a continued military presence and increasing tourism. Development on the other islands slowed after the war but has increased in recent years because of economic prosperity in Japan. Farming and cattle ranching have expanded on both Tinian and Rota. On Rota, a program of providing free homesteads to natives of the island has stimulated agricultural activities. As on Guam, tourism and associated development have expanded on Saipan and Rota because they are relatively close to Japan. Despite development, large areas of native and secondary forest remain on both Guam and Rota (Engbring and Ramsey 1984; Engbring and others 1986; Savidge 1984, 1988).

Although the Micronesian megapode (*Megapodius laparouse*) was extirpated from Guam at the turn of the century, the native avifauna of both islands that survived until the century remained largely intact through the 20th century until the 1960s. The resident avifauna of Guam included 28 species (17 land and water birds, 4 sea birds, and 7 introduced land birds). The resident avifauna of Rota included 23 species (11 native land and water birds, 6 sea birds, and 6 introduced land birds) as shown in Table 2-1. Since the middle of this century, Guam's avifauna has undergone a catastrophic decline, with the loss of 15 resident species (Table 2-1). The extinction of the Mariana mallard (*Anas oustaleti*) and

TABLE 2-1 Recent Resident Birds of Guam and Rota

Common Name (Scientific Name), Chamorro Name[a]	Guam	Rota
Mariana crow (*Corvus kubaryi*), aga	Rb	Rb
Black drongo (*Dicrurus macrocercus*)	I	I
Black francolin (*Francolinus francolinus*)	I	—
Blue-breasted quail (*Coturnix chinensis*)	I	—
Bridled white-eye (*Zosterops conspicillata*), nossa	Rx	Rb
Brown booby (*Sula leucogaster*)	Rx	Rb
Brown noddy (*Anous stolidus*)	Rx	Rb
Cardinal honeyeater (*Myzomela cardinalis*), egigi	Rx	Rb
Chestnut mannikin (*Lonchura molacca*)	I	I
Collared kingfisher (*Halcyon chloris*)	—	Rb
Common moorhen (*Gallinula chloropus*)	Rx	—[b]
Eurasian tree sparrow (*Passer montanus*)	I	I
Guam flycatcher (*Myiagra freycineti*)	Re	—
Guam rail (*Gallirallus owstoni*), koko	Re[b]	—
Island swiftlet (*Collocalia vanikorensis*), yayaguak	Rb	Rx
Mariana fruit-dove (*Ptilinopus roseicapilla*), totot	Rx	Rb
Marianas mallard (*Anas platyrhynchos oustaleti*)	Re[c]	—
Micronesian kingfisher (*Halcyon cinnamomina*), sihek	Re[d]	—
Micronesian megapod (*Megapodius laparouse*)	Re	R
Micronesian starling (*Aplonis opaca*). sali	Rb	Rb
Nightingale reed warbler (*Acrocephalus luscinia*)	Rx	—
Pacific reef heron (*Egretta sacra*)	Rb	Rb
Philippine turtle dove (*Streptopelia bitorquata*)	I	I
Red jungle fowl (*Gallus gallus*)	—	I
Red-footed booby (*Sula sula*)	—	Rb
Red-tailed tropic bird (*Phaethon rubricauda*)	—	Rb
Rock dove (*Columba livia*)	I	I
Rufous fantail (*Rhipidura rufifrons*), chichirika	Rx	Rb
White tern (*Gygis alba*)	Rb	Rb
White-browed rail (*Porzana cinerea*)	Rx	—
White-tailed tropic bird (*Phaethon lepturus*)	Rx	Rb
White-throated ground dove (*Gallicolumba xanthonura*), paluman apaka	Rx	Rb
Yellow bittern (*Ixobrychus sinensis*)	Rb	Rb

NOTES:
R = resident   I = introduced   — = absent   b = breeding   e = extinct   x = extirpated

[a]When known, the Chamorro name for the bird is included.
[b]Species extirpated from Rota during prehistoric times.
[c]Species has been seen on Saipan.
[d]Species survives in captivity.

Source: Information assembled from Engbring and others 1986; FWS 1990; and Reichel and Glass 1991.

extirpation of the white-browed rail (*Porzana cinereus*) during the 1960s were apparently caused by excessive hunting and the drainage and development of the island's wetlands (Engbring and Ramsey 1984; Savidge 1984). The loss of the remaining species during the late 1970s and 1980s has been related primarily to predation by the brown tree snake (Engbring and Fritts 1988; Reichel and others 1992; Savidge 1987, 1988). Both introduced and native birds have been affected (Conry 1987), and most species survive only in reduced numbers. In contrast, Rota's resident avifauna has remained relatively stable, with the loss of only a single species—the island swiftlet (*Collocalia vanikorensis*)—in the late 1970s (Engbring and others 1986). Even so, declines in the Rota populations of the bridled white-eye (*Zosterops conspicillata*) and rufous fantail (*Rhipidura rufifrons*) have been noted and are of special concern (Engbring and Pratt 1985; Engbring and Ramsey 1984; FWS 1981; Pratt and others 1979; Ralph and Sakai 1979). Predation by the introduced black drongo has been suggested as a principal cause of the decline of these small-bodied passerines (Craig and Taisacan 1994), but definitive studies have not been completed. The causes of this decline remain unknown. The apparent decline of the aga population on Rota is presumably associated with habitat loss related to development and perhaps other causes.

NATURAL HISTORY OF THE AGA

The aga is coal black, weighs about 250 g, and is about 38 cm long; males are slightly larger than females (Baker 1951). This crow has the lax plumage characteristic of island corvids (Goodwin 1986), with a slight greenish-black gloss to the head and back, bluish-black tail, and dull underparts (Baker 1951). Immature birds resemble adults but have less gloss and browner wings and tail.

The evolutionary relationship of the aga to other corvids is poorly understood. No genetic studies have been conducted on which to base a phylogeny. Possible founding ancestors could have come from Asia, Indonesia, Australia, New Guinea, or the Solomon Islands. On the basis of proximity and morphology, the slender-billed crow (*Corvus enca*) from Indonesia is probably its closest relative (Baker 1951; Goodwin 1986).

Aga are most abundant in mature forest (Baker 1951; Engbring and others 1986). Emergent trees are especially important as perch and nest sites on Guam (Morton 1996a). Avoidance of human habitation was noted by Baker (1951) and confirmed by recent studies (FWS 1990; Morton 1996a); crow sightings decline with increasing proximity to roads and aircraft runways (Morton 1996a).

Aga eat a variety of plants and animals. Principal foods are insects (grasshoppers, mole crickets, praying mantis, earwigs, seasonally abundant caterpillars [*Pericyma crugeri*], and lepidopteran larvae), small vertebrates (including immature rats), and invertebrates (hermit crabs, lizards, and bird eggs and nestlings), fruit, seeds, flowers, buds, foliage, and bark (Beaty 1967; Jenkins 1983; Lusk 1996, Michael 1987; Tomback 1986). They forage in the canopy and undergrowth and on the ground (Jenkins 1983; Tomback 1986). Information is lacking on seasonal variation in diet because most observations of foraging have been made during the breeding season.

Aga typically occur in pairs or groups of 2–5 (Jenkins 1983; Tomback 1986). Groups are most likely families consisting of a pair and their recently fledged offspring (Jenkins 1983; Tomback 1986). Occasionally, larger groups (of up to 14) are observed (Jenkins 1983). They appear to form monogamous pairs, but detailed studies of marked individuals have not been conducted.

A variety of vocalizations are used, including contact calls among group members during foraging (Jenkins 1983; Michael 1987; Tomback 1986). Pairs vocalize quietly at their nest with rambling "monologues" (Tomback 1986). Alarm calls are given in response to human activity and the presence of potential predators. Juveniles beg from their parents with squalling vocalizations (Morton 1996a; Tomback 1986).

Breeding can occur throughout the year, but a distinct peak occurs in the beginning of the dry season from October to December. Both parents construct the nest and care for the young (Jenkins 1983; Morton 1996a; Tomback 1986). Females do most of the incubation and brooding; males provision the females and nestlings. Incubation lasts 21–23 days (FWS 1990), and fledging occurs when nestlings are 36–39 days old (Morton 1996a). Clutch size is one to three eggs, but usually only one or two fledglings are produced (Morton 1996a). Reasons for nest failure (Table 2-2) include predation by brown tree snakes and monitor lizards (Aguon and others in press; Engbring and Fritts 1988; Jenkins 1983; Savidge 1987), nest abandonment due to unknown causes, egg infertility or nonviability (Aguon 1993; DAWR 1991, 1992; DFW 1990; Grout 1993; Wiles and others 1994), and disruption of breeding behavior by black drongos, other crows, and human activity (Grout 1993; Morton 1996a).

A typical aga nest consists of an outer platform of sticks (mean diameter, 37 cm) and an inner cup of vines, rootlets, and fiber (mean diameter and depth, 13 and 7 cm, respectively [Lusk and Taisacan 1996]). As with most crows, aga

TABLE 2-2 Fate of Nests on Guam After Development and Installation of Electric Barriers

Number of Nests	1992–93	1993–94	1994–95	1995–96
Found	28	40	27	8
Successful	0	2	0	2
Apparently abandoned	13	26	17	3
Apparently predated by snakes	4	0	1	0
Failed to hatch	3	4	7	1
Destroyed by researchers	0	2	0	1
Apparently failed for other reasons	3	1	2	1
Fate not known	5	5	0	0

commonly build and abandon multiple nest platforms before constructing a full nest, which can be constructed in as little as a week. New nests are constructed by a pair throughout a season after nest failure (Morton 1996a). Renesting 5 times within a season is not uncommon, and 7–10 successive nests have been documented (Morton 1996a). Prolific renesting by aga is likely an adaptation to their variable environment, where typhoons destroy many nests. It is an important characteristic that bodes well for restoration, in that more eggs could be harvested from a single pair of aga than from a pair of 'alala.

COMPARISON WITH THE 'ALALA

The general habits of the 'alala and aga are similar. Both occupy mature forest and forage on a variety of introduced and native plants, vertebrates, and invertebrates (Jenkins 1983; Sakai and others 1986, 1990; Giffin and others 1987; Morton 1996a). Both appear to be monogamous, but also form small family groups. Large groups have been documented in the aga (Jenkins 1983), and reported in the 'alala (Unger 1996).

As is the rule for corvids, male and female 'alalas and aga cooperate to raise a brood. Development to independence is slow in both species; parental care is extensive and can last up to a year (Banko 1996). Reproductive potential is much higher in the aga than in the 'alala. The 'alala rarely fledges more than 1 young from a nest and usually renests only once per year after nest failure (Banko and Banko 1980). The aga also typically fledges one young, but a pair will renest 5–10 times after nest failure (Morton 1996a).

Although both species are threatened with extinction, the reasons for their endangerment differ in important ways. Predation appears much more important as a limiting factor for the aga, whereas disease might be the primary limiting factor for the 'alala (Atkinson 1993; Jenkins and others 1989; vanRiper and others 1986). Food does not appear to limit either species, whereas senescence, inbreeding, and infertility might be affecting remnant populations of both. Aga appear slightly more gregarious than 'alalas.

DEMOGRAPHY OF THE AGA

This section describes what is known about the demography of the aga on the basis of analyses of data collected during field studies by the DAWR and FWS. Estimates of nesting success and adult survivorship are developed to help infer the causes of the population decline and for use in population projections.

Nesting success of the aga is different on Rota than on Guam. Although little field work has been conducted on Rota, John Morton (unpublished data) found 24 nests from 1993–1996. Of those 24, at least ten nests were known to fledge 14 young, for an average of 1.4 young per successful nest. One nest was destroyed by an unknown predator, but studies were not of sufficient intensity to be sure of the fate of ten nests that may have failed, and 3 others with eggs or chicks that were visited infrequently. Thus, a success rate of 48% (10 of 21 nests with known fates fledged young) should be considered a minimum for nesting success of the aga on Rota. Success rates for other species of corvids vary considerably (mean = 58%, standard deviation = 19 for 15 species; NRC 1992) and the aga on Rota is within this normal range.

On Guam, nesting success is extremely low and below that observed in other corvids. Since 1986 only five aga nests have been known to fledge young, and all were on nest trees protected with electric barrriers. Post-fledging survival is poorly known, although the single chick fledged in 1989 apparently was killed soon after leaving the nest by a monitor lizard (Aguon and others in press). Before the use of electric barriers to protect nests, most nest failure was probably due to predation of eggs and nestlings by the brown tree snake, or to nest abandonment prior to or during egg-laying, perhaps as a result of harassment of nesting females by the snake. Exact causes of nest failure are often difficult to determine because direct evidence of the cause of death is often lacking. For example, many empty crow nests are found apparently abandoned (Table 2-2), and the cause of their failure is often unknown. In addition to predation by the brown tree snake, nest failure is also apparently caused by typhoons, local storms, and predation by monitor lizards.

The installation of electric barriers on nest trees has helped to decrease the loss of eggs to the brown tree snake, but these barriers have not resulted in greatly increased reproductive success (Table 2-2). Apparent abandonment of the nest occurred at 59 (63%) of the 93 nests with known fates and was the dominant form of nest failure. Failure of eggs to hatch was the second greatest cause of nest failure, occurring at 15 (16%) of the nests with known fate. Even protected nests that were not abandoned rarely produced crow young. Of 42 eggs that remained in the nest through incubation, 35 (83%) failed to hatch (Table 2-3). Many eggs showed no development and appeared to be infertile, while others were intact or were cracked and contained dead embryos (Brock 1996). Lack of development or embryo mortality could be due to the effects of advanced age of parent birds that has resulted from at least a decade of low recruitment, to poor incubation of eggs caused by the harassment of incubating females by snakes, or to other, unknown causes.

TABLE 2-3 Egg-Hatching Success on Guam After Development and Installation of Electric Barriers

Number of Eggs	1992–93	1993–94	1994–95	1995–96
Laid	12	17	12	9
Hatched	0	3	1	3
Failed to hatch	6	12	11	6
Lost to other causes	6	2	0	0

Changes of reproductive function with age are not well known in wild populations of birds (Holmes and Austad 1995; Martin 1995). When it does occur, it can take many forms. Effects of age may be manifested in the form of smaller or even larger clutch sizes, fewer or more clutches per year, earlier dates of first breeding, and lower or higher frequency of breeding and renesting (Austad 1995; Clutton-Brock 1988; Fowler 1995; Holmes and Martin 1995; Sæther 1990). The effects of senescence often are not expressed uniformly in a population, but vary greatly among individuals. Although fertility can decrease with age, the effects of reproductive senescence usually differ for individuals within a population and decreases are less likely in pairs with long term pair bonds (Clum 1995; Fowler 1995).

Egg infertility, however, is difficult to distinguish from early embryo deaths due to other causes. Early embryo death might be caused by inadequate incubation by a female who was being harassed by snakes. The viability of developing eggs declines with exposure to temperatures above 26–27°C (but below optimal incubation temperatures (34–36°C) (Romanoff and Romanoff 1972; Webb 1987; White and Kinney 1974). This effect is stronger in tropical environments, where temperatures often exceed 26–27°C but are below optimal incubation temperatures for most of the day (Stoleson and Beissinger 1995; Stoleson 1996; Veiga 1992). Apparent egg infertility could also be the result of eggs overheating (Webb 1987). If females left their eggs exposed to the sun due to disturbances or other causes, the embryos might die from overheating. In addition, unknown causes could also be responsible for the low hatchability of eggs. Behavioral studies of egg-laying and incubating females are needed to help distinguish possible causes for the failure of eggs to hatch.

In addition to low nesting success, productivity for the Guam population has been low because many pairs did not lay eggs at all (Table 2-4). Only one-third to one-half of the territorial pairs laid clutches in any year. Those proportions may be underestimated because some pairs might have laid eggs that were eaten by snakes before they were detected by researchers, or might have abandoned nests when electrical barriers were installed. However, it is clear that nonlaying pairs composed a substantial proportion of the aga population on Guam.

TABLE 2-4 Recent History of Nest Initiations on Guam.

Number of	1992–93	1993–94	1994–95	1995–96
Pairs that nested	12	11	9	6
Pairs that laid clutches	5	5	4	2
Nests found	28	40	27	8
Nests receiving clutches	9	7	8	4
Nests without clutches	19	33	19	4
Young fledged	0	2	0	2[a]

[a]Young raised from wild eggs and hand-raised in captivity.

The conventional measure of productivity of a bird population is the number of young fledged per pair. In Guam, only six young are known to have fledged from 1989 through 1996, all in nests that were protected by barriers. Even during the past four years when nearly every nest was protected with electric barriers, only four young have been produced, and two of these were hatched and hand-raised in captivity in 1995–1996 (Table 2-4). This low level of productivity could not sustain a population of crows or any other bird species—survival would have to be near 100% to support a population with such low productivity. Such high survival is not normal in wild bird populations.

No studies of the survival of marked birds have been conducted on either Rota or Guam. On Guam, but not Rota, several kinds of information can be used to estimate adult survivorship indirectly. Crows have high fidelity to territories and even single birds remain on territories for years. Rates of territory turnover (Table 2-5), defined as the loss of 1 or more members of a territorial pair, can be used as a surrogate measure of survival. Here we estimate survivorship from territory turnover rates in two ways:

- If female aga, which are thought to be the only sex that incubates eggs, are susceptible to predation by brown tree snakes when they are incubating, a single bird lost from a territory is perhaps likely to be a female. The possibility of a male-biased sex ratio seems likely because several other species of forest birds on Guam had male-biased sex ratios just prior to their extinction, and because groups

TABLE 2-5 Number of Aga Occupying Territories on Guam

		Number of Birds[b]		
Location	Pair Designation[a]	1993–94	1994–95	1995–96
Conventional-Weapons Storage Area	Mag 8 OL	1	0	0
	New Mag OL	0	0	2
	4th St.	2	1	0
	15B3	2	2	2
	C1	2	2	2–1
	11B6	1	2	2
	C4	2	1	0
	Borrow Pit	2	1	0
Pati Point	Pati Point	1	1	1
	Late	2	0	0
	Tarague Jeep Rd.	2	2	2
	Pipeline	2	2	1
	Tagua	2	2	2
	WSA	2	2	2
	Breadfruit	2	2	2
Northwest Field	Area 10	2	2–1–2	0

[a]Nest-site designations assigned by DAWR.
[b]Changes in number indicate loss or gain of individuals.

of presumed-male aga have been observed pursuing and harassing nesting females in recent years (Beck and Aguon 1996). A measure of annual mortality of females can be estimated from the proportion of territories that lose at least one pair member from year to year. The aga breeding season begins in October. From late 1993 until late 1994, 4 of 12 territories (33%) lost one or more pair members, and from late 1994 to late 1995, 2 of 9 territories (22%) did so (Table 2-6). The turnover rate for the two years combined was 29% (6 of 21 cases). Average annual survivorship is then 100 minus 29, or 71%.

- If deaths are equally distributed between the sexes, their joint survivorship can be estimated by the proportion of territorial birds remaining over time. This approach assumes that birds will remain on their territory even when they have lost their mates. Data are available for territorial pairs for two years beginning in the 1993–1994 breeding season (Table 2-5), and the fate of the original 27 birds can be traced (Table 2-6). Only 15 of the 27 crows (56%) were accounted for two years later. The estimate of annual survivorship is the square root of the proportion of birds remaining after two years (0.56) and this yields an estimate of 75%.

Both of the above estimates of adult survival (71% for females and 75% for both sexes combined) are lower than those of other birds on islands of similar size. Most long-lived birds, like crows, have annual survivorship between 80–90%. Annual survivor ship of the 'alala was estimated at 81–90% (Banko and Banko 1980; NRC 1992) on the McCandless Ranch where the population was stable, but only 66–70% on Honaunau and Haulalai, where populations have declined greatly (NRC 1992) and now are extinct.

TABLE 2-6 Population Structure and Territory Turnover on Guam from 1993 to 1996[a]

Population Structure and Tterritory Turnover	Start of Breeding Year		
	1993–94	1994–95	1995–96
Structure			
Total number of birds	27	22	18
Total territories occupied	15	13	10
Total pairs	12	9	8
Total singletons	3	4	2
Percentage singletons of total population	1.1	18.2	11.1
Percentage territories occupied by singletons	20	30.8	20
Turnover			
Pairs with turnover	—	4	2
Percentage pairs with turnover	—	33.3	22.2
New pairs formed	—	1	1
Singletons lost	—	1	3

[a]Based on information in Table 2-5.

POSSIBLE CAUSES OF POPULATION DECLINES

The aga population on Guam has decreased progressively over a period spanning nearly 30 years. Although the crow's decrease in distribution has paralleled that observed in Guam's other native forest birds (Jenkins 1979, 1983; Savidge 1984), it has been less rapid, extending over a longer time. On Rota, decreases in the crow's distribution and abundance have occurred more recently and have been much less severe. Various factors—including habitat loss and modification, human exploitation, typhoons, pesticides, competition, disease, predation, food resources, and, more recently, inbreeding and senescence—have been proposed as causal agents for the decrease. Each factor is discussed briefly below.

Habitat Loss and Modification

The specific habitat requirements of the aga are not well known, but available information suggests that the crow requires relatively undisturbed native forest (Baker 1951; Engbring and others 1986; Marshall 1949; Seale 1901; Strophlet 1946; Tomback 1986). The effects of World War II on Rota were considerably less than on Guam, and the habitat changes associated with economic and commercial development have been more recent (Engbring and others 1986). Although those differences suggest that the declines of the crow populations on both islands are related directly to habitat loss and modification, such a conclusion is difficult to support, considering that substantial tracts of native forest remain in both southern and northern Guam yet are unoccupied by crows (FWS 1990; Savidge 1984, 1987); that most avian populations on Rota, Saipan, and Tinian have remained relatively stable despite extensive habitat alterations (Engbring and others 1986; Savidge 1987); and that some pairs of crows on Rota have continued to occupy their territories despite relatively extensive habitat modification (Grout 1996). These observations are more easily reconciled, however, if one assumes that the causes of decline operating on each island have been different and recognizes that established pairs of aga, like other corvids, will sometimes remain on their territories despite considerable habitat alteration. Nevertheless, continued habitat modification and fragmentation will ultimately compromise the long-term survival of the species on either island.

Human Exploitation

Although most forest birds on Guam have been legally protected since the turn of the century, the aga was not protected until 1981 (FWS 1990; Savidge, 1988) and has traditionally been considered an agricultural pest (Safford 1902; Seale 1901). Furthermore, according to Beaty (1967), local Chamorro superstition held that illness would follow hearing a crow call. Although such cultural prejudices undoubtedly have resulted in indiscriminate killing of crows in the past, there is no evidence that such killing was responsible for the crow's recent population decline on Guam. Since the early 1980s, the extant crow population has been restricted largely to military lands, where the hunting of native species is prohibited and tight security has minimized poaching (Savidge 1988).

On Rota, however, some illegal killing might continue. In at least one documented instance, a nesting pair was killed on forest lands that were being cleared for commercial development (DFW 1993; Grout 1996). Given the increasing pressures for development of private lands on Rota and the negative attitudes toward protection of endangered species, the crow's endangered status itself has probably contributed to illegal killing.

Typhoons

Major typhoons have struck Guam and Rota in recent years, and, as previously mentioned, the destruction of crow nests and deaths of nestlings have been documented on both islands (DAWR 1991, 1992; DFW 1990; Grout 1993; Wiles and others 1994). Unfortunately, the effects of these storms on the native avifauna, including the crow, have not been well documented (FWS 1990). On Saipan, native birds remain abundant despite several major typhoons during the last 25 years (Engbring and others 1986; FWS 1990). Recent studies in the Caribbean have documented that such storms can substantially damage and modify native forests and food supplies (Bellingham and others 1995; Scatena and Larsen 1991; Walker 1991) and result in considerable death and dispersal, particularly in species that depend on constant supplies of fruit and nectar (Askins and Ewert 1991; Lynch 1991; Pierson and others 1996; Waide 1991; Wauer and Wunderle 1992; Wiley and Wunderle 1994; Will 1991; Wunderle 1995; Wunderle and others 1992). Although native birds in the Mariana Islands have evolved with the periodic occurrence of major typhoons, the effects of typhoons on small, remnant populations could be especially severe. For example, Hurricane Hugo in 1980 resulted in a 50% reduction (from 47 to 23) in the population of the endangered Puerto Rican parrot (*Amazona vittata*) (Collar and others 1992). Increased habitat modification and fragmentation will certainly reduce the probability that insular populations will survive and recover quickly from such storms.

Pesticides

From World War II until the early 1970s, various pesticides, including DDT and malathion, were used extensively on Guam for vector control and agricultural purposes (Baker 1946; FWS 1990; Savidge 1984, 1987). However, a 1981 survey of pesticides in guano, birds, and small mammals indicated that it was unlikely that pesticides were responsible for the continuing decline of Guam's avifauna (Grue 1981). And in some areas, the decline did not occur until many years after the ban on DDT. The history of pesticide use on Rota is unknown, but various insecticides have been used regularly and are readily available. Although there is no indication that the crows on either island have been adversely affected by pesticides, no thorough studies have been completed.

Competition

As noted above, seven species of land birds have been introduced to Guam, and six have been introduced to Rota (Reichel and Glass 1991). Of these, the only species that could be considered to be a potential competitor is the largely insectivorous black drongo. The drongo was introduced to Rota in the 1930s and spread naturally to Guam in 1960 (Savidge 1987). The density of black drongos has remained lower on Guam than on Rota, presumably because of the high rate of nest predation by the brown tree snake on Guam (Beck 1996). Maben (1982) concluded that drongos were not competing directly with native species, including the crow, because of differences in habitat use and foraging techniques. However, drongos have been seen harassing nesting crows (Grout 1996). Differences in nest placement by crows on Rota (in the subcanopy) and Guam (in the canopy) have been ascribed to differences in drongo densities, and hence harassment rates, between islands (Tomback 1986). Additional effects of such interactions on breeding and incubating crows remain unknown.

Disease

Extensive disease and sentinel studies (studies where data on animals exposed to contaminants in the environment are regularly and systematically collected and analyzed to identify potential health hazards to other animals or humans [NRC 1991]) conducted on Guam in 1982–1984 did not reveal any evidence that the declines in the native avifauna were caused by infectious disease (Savidge 1987). Although no similar detailed studies have been conducted on Rota, health examinations on captured crows, bridled white-eyes, and Mariana fruit doves have not revealed pathogenic organisms or parasites (Olsen 1996). Postmortem examinations have been done opportunistically in the past. A more concerted effort needs to be made to recover all dead embryos, chicks, or fledglings for complete postmortem examinations. Pathology protocols need to be developed in advance and applied consistently in order to maximize the usefulness of the data. Having accurate, comprehensive pathology data will help answer questions regarding causes of mortality in various age classes and the significance of any subclinical infections or nutritional problems.

Predation

Various predators of birds have been introduced to the Mariana Islands, including dogs, cats, rats, monitor lizards, and the brown tree snake. All except the brown tree snake are present on all the major islands of the Mariana archipelago and have coexisted with the native avifauna for a long time without causing important population declines or extinctions. As discussed below, the population declines of Guam's native avifauna, including the crow, appear to be caused directly by predation by the extremely numerous brown tree snake (Engbring and Fritts 1988; Reichel and others 1992; Savidge 1987). Because the brown tree snake has now been detected repeatedly on Saipan and other Pacific islands (Fritts 1988), a series of unintended experiments are now under way.

Food Resources

Recent increases in the number of nonlaying pairs and in the production of nonviable eggs in the Guam aga population could possibly be related to food or micronutrient resources and foraging energetics. Clearly, the brown tree snake has dramatically altered the fauna of Guam. Some food resources previously used by the aga, such as the eggs and nestlings of other forest birds, have been eliminated by the snake. Other important resources, such as skinks and geckos, survive, but at substantially reduced population levels because of snake predation. A recent study of lizard populations on 10^2-m plots with snakes (36 snakes/ha) and without snakes indicated that overall skink and gecko densities and biomass were about 33% less with predation by snakes (density, 13,210 vs 19,650 lizards/ha; biomass, 35 vs 53 kg/ha). Although those differences are substantial, there still seem to be sufficient prey for the aga. Furthermore, given the wide variety of food items regularly consumed by crows, it seems unlikely that food resources could be limiting the population, except perhaps for brief periods after typhoons. However, because the brown tree snake does not appear to occur on Rota, the food- shortage hypothesis could easily be tested by comparing the rate at which males feed their mates on the nest on the 2 islands.

Inbreeding

Because the Guam population of aga is small and has been declining for years, it might be hypothesized that reduced productivity is the result of inbreeding depression (Ralls and others 1980, 1986; Schoenwald-Cox and others 1983; Soulé 1987). That possibility seems slight, however, considering that although the genetic diversity of the Guam population is greater than that of the Rota population, the Rota population does not appear to be experiencing any reductions in egg fertility, egg hatchability, or chick survival. Again, comparative information is needed to test this hypothesis.

Senescence

In the last few years, breeding pairs of crows on Guam have produced a number of infertile or nonviable eggs (nonviability of eggs can be caused by, for example, infertility, inbreeding depression, thin or cracked egg shells, poor parental nutrition, or inadequate incubation) or have constructed nests that lacked eggs when examined. That has resulted in the suggestion that many of the crows in the Guam population are exhibiting reproductive senescence—a suggestion based on the premises that the population has been progressively declining for a number of years, that recruitment to the population has been extremely low, and that most of the birds in the population are reaching the end of their life span. Declines in reproductive performance in older birds have been documented in a number of species (Clutton-Brock 1988; Holmes and Austad 1995; Martin 1995; Saether 1990). However, senescence can be manifested in a variety of forms (such as egg fertility, clutch size, laying dates, and renesting potential), and these effects usually vary greatly between individuals (Clum 1995; Fowler 1995).

Comparative information on the black-billed magpie (*Pica pica*) (Reese and Kadlec 1985) and the rook (*Corvus frugilegus*) (Røskaft and others 1983) indicates that older females typically are more vigorous breeders, laying larger clutches earlier in the breeding season, hatching more nestlings, and raising more fledglings than younger females. However, both male and female Florida scrub jays show a drop in survival and reproductive output with extreme age, although a great deal of variation exists between individuals. Some females continue laying until their death; others lay smaller clutches, are less likely to renest, experience slightly decreased hatching success, lay mutant eggs, or cease laying altogether but act territorial and build nests (Fitzpatrick 1996). Similar patterns (egg infertility, reduced renesting, and nest construction without egg-laying) have been observed in the small remnant wild population of 'alala. However, as noted by Clum (1995) and Cézilly and Nager (1996), in species with long-lasting pair bonds, individuals might enhance breeding performance by repeatedly breeding with the same partner; hence, an apparent effect of senescence could result partly from an overrepresentation of "new" pairs in older age classes. Few individuals in the Guam population are banded, so it is difficult to discriminate between the alternative hypotheses. Because of the difficulties in monitoring nests in the field, it is also possible that some cases of "egg infertility" actually represent early embryo deaths brought on by poor incubation by adults as a result of disturbance by snakes. Similarly, cases of "nonlaying" might simply represent snake predation of eggs soon after laying.

Summary

In summary, the principal cause of the decline of the aga population on Guam appears to be predation by the brown tree snake, although the reproductive effects of advanced age, frequent remating due to natural deaths, nest abandonment, or some as-yet-unidentified factor cannot be ruled out as a more recent problem. On Rota, the decline of the crow population appears to be related primarily to habitat modification and secondarily to human-caused deaths associated with land development. The brown tree snake might be present on Rota, but as yet it has not been reported there.

POPULATION PROJECTION FOR THE AGA ON GUAM

The committee was asked to conduct a population viability analysis (PVA) for the aga. Such analyses usually use stochastic models to make long-term projections of populations and to make probabilistic estimates of the rate of extinction under different management options (Burgman and others 1993). A PVA can predict the size of a population for 50–200 years by allowing the rates of fecundity and survivorship to vary stochastically for each year, based on estimates of the variance in these parameters. This requires good estimates of the annual variation in survival and fecundity that results from demographic stochasticity, environmental stochasticity and catastrophic events (Boyce 1992). At least 10 years of demographic data are required to calculate valid variances in fecundity and survivorship for a long-lived bird like the aga. Such data are not available for the aga, as the review of its demography in the previous section indicates.

Even when parameter values are available, the structure of PVA models may be inappropriate and hence the results may be unreliable (Caughley 1994; Caughley and Gunn 1995). For example, PVA models hardly ever consider the interaction between the target population of interest and the population dynamics of its predator. For the aga, this would be a serious shortcoming, since changes in the dynamics of the brown tree snake could have a large effect on the population dynamics of the aga. Thus, the committee did not construct a PVA model for the aga because it was inappropriate.

The committee considered using a simplified Leslie-Lefkovitch matrix to predict lambda, the short-term rate of annual change in the population. The Leslie-Lefkovitch matrix has been a standard tool to evaluate population dynamics for more than 50 years (Leslie 1945; Lefkovitch 1965). However, the same concerns that led to the realization that a PVA analysis is inappropriate apply equally well to lambda. Furthermore, it is doubtful that such an analysis would lead to a different conclusion than that which is strongly indicated by the evidence, namely, that the aga population on Guam is in extreme danger of extinction and that the major cause is adult mortality. While more sophisticated analyses might be useful in the future, once recovery is underway, they would require much better, more reliable data.

PAST AND CURRENT CONSERVATION AND MANAGEMENT

The major government organizations that have been involved with conservation and management of the aga on Guam and Rota have been DAWR, DFW of the CNMI Department of Lands and Natural Resources, FWS, and the National

Biological Service (NBS) of the US Department of the Interior (DOI). The US Air Force and Navy, the US Department of Agriculture, and the DOI Office of Insular Affairs (formerly the Office of Territorial and International Affairs) have played smaller roles. Since 1993, the American Zoo and Aquarium Association and 3 member zoos have participated in the Marianas Archipelago Rescue and Survey (MARS) program's efforts to add the aga to their captive-propagation activities. With the implementation of an intensive management plan for the aga by DAWR in 1994, there has been increasing interaction with the 'alala recovery program in Hawaii. All those participants share one or more of the following objectives: to diagnose the causes of the population declines in the aga on Guam and Rota, to ensure that it survives in the wild, and to eradicate or at least control the brown tree snake on Guam and ensure that it does not spread to other islands.

In 1981, the aga received legal protection by the government of Guam for the first time; and in 1984, it was federally designated as endangered (FWS 1984). Of the 6 other endangered species listed at that time, 2—the Mariana moorhen and the island swiftlet—still survive. The Guam broadbill and the Guam bridled white-eye were extirpated from Guam in the 1980s, and the Guam rail and the Guam Micronesian kingfisher exist now only in captivity. A recovery plan for endangered birds of Guam and Rota prepared by Robert Beck and Julie Savidge (FWS 1990) recommended a captive-breeding program for the crow and study of the feasibility of translocating birds from Rota to Guam. In 1991, the crow was among the Guam birds proposed for the designation of critical habitat by FWS (1991). The proposal was withdrawn (FWS 1994) because, with the inclusion of these lands in the newly established Guam National Wildlife Refuge, most of the lands proposed for critical habitat became protected. The refuge consists of about 150 ha of newly protected areas and more than 9,000 ha of current military lands at the northern end of Guam. The Pacific Island Recovery Team (PIRT) in FWS is charged with developing a recovery plan for the aga and the other endangered birds of the Pacific islands.

Guam

The decline in the aga population on Guam has been documented with stationary counts, roadside counts, and sampling along transects. Roadside counts conducted by DAWR since the 1960s showed that the aga had disappeared from southern and central Guam by the early 1970s (Aguon 1983; Jenkins 1983). In 1981, Engbring and Ramsey designated randomly selected starting points and parallel transects and used a circular-plot-survey method (Reynolds and others 1980) to estimate the number of crows on Guam. After correction for detection distance and differences among observers, their record of 241 birds led to a total population estimate of 357 (Engbring and Ramsey 1984). By the middle 1980s, it was becoming apparent to researchers that although there were many threats to the persistence of the aga and other native birds on Guam, the overriding threat was predation on eggs and young birds by extremely high populations of the brown tree snake (Conry 1988; Engbring and Fritts 1988; Savidge 1986, 1987). A repetition in 1990 of Engbring and Ramsey's 1981 transect survey by DAWR recorded only 107 birds. This survey used recorded-call playbacks, which are about twice as effective a detection method as the standard circular-plot detections used in 1981. The last record of unassisted fledging of juvenile crows was in 1985 (Michael 1987). DAWR estimates of the number of crows on Guam fell from 100 in the breeding season of 1990–1991 to 18 territorial birds and a few nonterritorial birds in 1995–1996 (Anderson 1996). Only 1 of the pairs of the crows remaining in 1995–1996 produced eggs.

DAWR has developed an electric and mechanical barrier, tested it on trees, and shown that it is potentially effective in excluding snakes and monitor lizards from nest trees of the aga (Aguon and others in press) and improving the ability of the birds to nest (Aguon and others 1992). The development of the tree-barrier technology began in 1985; by 1989, it was combined with trapping snakes in and around the barriered trees. In 1989, a chick fledged from a nest protected by a tree barrier but died soon thereafter. Improvements in tree-barrier technology led to the fledging of three chicks in 1991–1992. The finding that females were laying larger clutches (3 versus 1–2 eggs) in nests in protected trees during the 1992–1993 breeding season strongly supported the suggestion that in previous years many eggs were lost to brown tree snakes during the laying period.

In 1993–1994, DAWR researchers discovered that a high percentage of eggs laid were not viable, so an intensive aviculture intervention program was begun to study this phenomenon, to incubate eggs artificially, and to hand-rear chicks. The crows readily renest, so taking eggs ("double clutching") encourages them to produce more. One chick was successfully hatched in 1994–1995 and was returned to its nest 2 days after hatching, but it died before fledging. In 1995–1996, although only 1 pair of crows on Guam laid fertile eggs, 1 male and 1 female aga were successfully reared from their eggs. These young birds remain (August 1996) in captivity on Guam, but their wild male parent has disappeared.

Simultaneously with the research efforts for the aga on Guam in the 1980s, an intensive research program on the brown tree snake was begun by Savidge (1986, 1987). From necropsies of snakes, she determined that in the period 1982–1986, a time when birds had virtually disappeared from much of Guam, the percentage of birds and bird eggs in the diets of snakes was lower than it had been in the 1970s, and the percentage of lizards was higher. By using sentinel birds (coturnix quail, domestic chickens, domestic canaries, and wild-caught bridled white-eyes from Saipan) placed in the forest, she determined that predation on eggs and young birds was high. There was no evidence that either disease or chemical contamination was causing declines in the birds (Grue 1985; Savidge 1986). Continuing efforts by NBS (efforts formerly housed in FWS) have greatly extended the research on snakes (Fritts 1988; Rodda and others 1992, in press a). Field studies in the early 1990s of the behavior of the aga on Guam during the breeding season showed that many nests were abandoned before they were completed and that many eggs failed to hatch. One must take into account that corvids readily renest. But other factors might contribute to the inability of the birds to complete the nesting cycle, such as harassment by unmated aga, disturbance during low overflights by aircraft, disturbance by snakes, and infertility of eggs (DAWR 1993; Grout 1993; Morton 1996a; Wiles and others 1994). Even full protection from the brown tree snake might be insufficient to keep the aga from being extirpated from Guam.

In 1994, DAWR implemented an aviculture intervention plan in which eggs laid in the wild would be taken for artificial incubation. Captive hatchlings would be hand-reared and then either placed back in nests that had been held in the incubation phase with dummy eggs or released as juveniles. Protocols for this project were developed in consultation with the personnel of the 'alala recovery program in Hawaii. After a series of problems with infertility and incomplete development, the 2 captive birds mentioned above fledged in captivity in March 1996. In late 1995, DAWR requested a FWS permit to translocate eggs and young aga from Rota to Guam. Because FWS anticipated public and scientific interest in the proposal to introduce birds from Rota to Guam, and because comments it received showed concern about such actions, FWS requested this NRC study. Fulfillment of the permit request awaits the findings of this committee.

The need for herpetologic expertise and snake-control activities applicable to the protection of the aga within the DAWR program is clear. Even so, there are vacant positions in the organization and sizable amounts of unexpended funds earmarked for such work. The staff herpetology position, which was vacated in 1992, remains unfilled. In summary, conservation and management efforts for the aga on Guam have emphasized surveys, behavioral observations of unbanded birds, and placement of tree barriers to protect nests from predation by brown tree snakes and monitor lizards. Most recently, they have included intensive management efforts involving the removal of eggs from wild nests for artificial incubation and hand-rearing of the young. In spite of those efforts and the protection of crucial forested habitat as part of the Guam National Wildlife Refuge, the status of the current wild population of the aga on Guam must be regarded as desperate.

Rota

Field work by DFW on Rota has been limited by a lack of permanent staff. Field observations on Rota indicate that the aga there forages and nests within the forest canopy and is less conspicuous than on Guam. There is some information on the omnivorous diet (which includes fruits and insects) and behavior of the aga on Rota (Tomback 1986). Evidence of an inability to complete construction of nests or the nesting cycle has been described in annual CNMI reports in the 1990s and is reminiscent of similar observations on Guam. However, one must take into account the tendency for corvids to build multiple platforms before completely constructing a nest. Several chicks on Rota have been banded, but there have been no consistent followup efforts (Grout 1996; Morton 1996b).

There have been differences of opinion about the size and status of the Rota population of crows. Pratt and others (1979) considered the species to be common in spite of persecution by hunters, whereas Ralph and Sakai (1979) considered it to be uncommon. In 1982, Engbring and others (1986), using randomly selected parallel transects and a circular-plot-survey method, recorded 454 crows on Rota. After correcting for an estimated detection distance of 123 m and for observer effects, they estimated that there were about 1,300 crows on Rota. Because this survey method involves estimating detection distances and extrapolating numbers for unsurveyed areas, it undoubtedly is less reliable for crows than other potential methods. Similar procedures in 1994, as part of a contracted 5-year population survey, produced a population estimate of 592 birds (95% confidence interval, 474–720). The reliability of those estimates is difficult to judge; the aga on Rota is probably declining together with the bridled white-eye and rufous fantail. The magnitude and causes of these declines are not understood, although obvious evidence of predators is lacking.

The most recent population estimate of 592 crows on Rota is based on a variable circular-plot survey in October and November 1995. The survey used the same methods and many of the same transects as were used by Engbring and others (1986) and was conducted in the same fashion. Grout and others (1996) reanalyzed the 1982 survey data of Engbring and others and revised the total estimate from 1,318 to 1,348. Then, comparing that estimate with the estimate for 1995, Grout and others (1996) estimated that the total population had declined by 50% between 1982 and 1995. If one assumes that the 1982 transects that were not repeated had no crows, a direct comparison of the numbers of crows observed on the same 11 transects that were repeated plus the 2 that were not shows a much greater decline—about 71%—in 13 years (from 224 to 64).

DFW is convinced that there have been substantial declines in the Rota population of crows in the last decade (Palacios 1996). Seventeen known territories of crows have been mapped in 1994–1996 (Grout and others 1996). In 1994, FWS opened an office on Rota. A contractor (Resources Northwest, Inc., Bellevue, WA) has been hired on Rota to develop a habitat-conservation plan for the island (see chapter 5, "Habitat Protection").

The 2 major concerns on Rota are the danger that the brown tree snake will or may already have reached the island in cargo arriving by ship or air from Guam and that insufficient forest habitat will be protected from development to sustain populations of forest birds.

A captive population of aga was founded with wild birds captured on Rota in 1993–1995 as part of the MARS program. The MARS program includes 7 zoologic institutions (Houston Zoological Gardens, National Zoological Park, Louisville Zoological Garden, Memphis Zoological Garden and Aquarium, Honolulu Zoo, North Carolina Zoological Park, and Philadelphia Zoological Garden). The program was initiated largely to develop techniques for the capture, acclimation, transport, and propagation of aga, Rota bridled white-eyes, and Mariana fruit doves; to provide training to FWS, DAWR, and DFW biologists in field capture and acclimation techniques; and to conduct basic research on the past and present distribution of Mariana birds.

To date, 10 wild aga have been collected from Rota: 7 in July 1993 as part of the first MARS collection, 1 in July 1994 by CNMI biologist David Worthington, and 2 in January 1995 as part of the second MARS collection. Of the 7 birds collected in 1993, 2 (1 male and 1 female) were sent to the Philadelphia Zoological Garden for quarantine and then were transferred to the Houston Zoological Gardens (HZG) for breeding. The remaining 5 (3 males and 2 females) were sent to the National Zoological Park's Conservation and Research Center (NZP–CRC) in Front Royal, VA. The single female salvaged in July 1993 was in ill health at the time of capture and was given antibiotic therapy on Rota before being transferred to NZP–CRC later in the same month. She died on January 9, 1996, despite continuous medical treatment throughout her captivity. The necropsy revealed that death resulted from renal failure. Although the cause of the condition was not determined, the persistent high blood concentrations of uric acid obtained from the time of capture suggest that poor renal function was the underlying cause of her original illness in the wild. The last 2 birds (1 male and 1 female) in the captive population were collected on Rota in January 1995 and transferred to NZP–CRC.

Two offspring were hatched by the pair of crows at the HZG on April 21 and 22, 1995. The second nestling disappeared from the nest at the age of 5 days and presumably was eaten by the adults, inasmuch as no remains could be found in the nest or elsewhere in the breeding enclosure. The first nestling (a female) is still housed at the HZG. Although 3 adult pairs held at NZP–CRC have laid eggs, no offspring have as yet been produced. All clutches have been either infertile or destroyed by the adults during incubation.

Habitat Protection

As mentioned above, about 9,000 ha of forested habitat on Guam was recently incorporated into the Guam National Wildlife Refuge for the preservation, protection, and management of endemic endangered birds (FWS 1994). In addition to these refuge lands and private holdings, the government of Guam owns and manages some 1,620 ha of suitable forest habitat (FWS 1994). This amount of land—if kept in its current condition—should be adequate for maintaining a viable population of aga on Guam. Ideally, however, additional forest lands at various locations throughout the island should be acquired to minimize the potential risks associated with catastrophic events, such as typhoons.

On Rota, Engbring and others (1986) reported that native forests made up about 60% (5,100 ha) of the island's area in 1982. The total has probably decreased by about 5–10% in the intervening years (Grout and Lusk 1996). With increasing development pressures on Rota, the loss and fragmentation of native forest habitat is almost certain to accelerate in the immediate future. Because most of the land on Rota is privately owned, a habitat-conservation plan for the island is being developed to preserve essential forested habitat for the aga and other native wildlife. FWS has assigned a full-time biologist on Rota to assist in information-gathering and planning and is providing geographic information system (GIS) support to CNMI planners (Grout and Lusk 1996).

Habitat conservation plans offer a mechanism for creative partnerships between the private sector and local, state, and federal agencies in protecting and preserving endangered species and their ecosystems under the Endangered Species Act. However, this approach has been used infrequently. It has sometimes been impossible to achieve consensus on a plan, but where such consensus has been achieved and plans have been implemented, very promising results have been seen (NRC 1995). Because development of the Rota habitat-conservation plan has only recently begun to be implemented, it is difficult to evaluate its eventual role in the conservation of the Rota population of the aga. Ultimately, however, the success or failure of the plan will depend on whether enough native forest is set aside for the maintenance of a viable crow population. Several wildlife-conservation areas have already been established on Rota (FWS 1996), but they are not inhabited by crows.

The biologists on both Guam and Rota and the various workers from FWS and academe who have worked there are dedicated to protection of the endangered native biota, but the higher levels of government and the general public consider economic development to have a higher priority. This poses a potential conflict with crow-habitat preservation (Kuhlman 1996.) An approved islandwide habitat-conservation plan for Rota might protect 3 areas as forest but allow development of resorts and homesteads in some areas that are now occupied by crows. The objective is to allow economic development in concert with protection of native plants and animals. In spite of previous work on both Guam and Rota (Jenkins 1983; Marshall 1949; Tomback 1986; annual reports of DAWR and DFW), the life history, social structure, and ecology of the aga are not fully understood, so a full diagnosis of all the factors that might be contributing to its decline is not yet possible. There is an urgent need for new intensive programs of research and management for both the aga and the brown tree snake.

3

Brown Tree Snake

NATURAL HISTORY

The brown tree snake, *Boiga irregularis*, like other members of the genus *Boiga*, is a slender, arboreal, nocturnal snake with grooved venom fangs at the posterior margin of the maxilla. It subdues prey by biting, constriction, and injection of a mildly toxic venom. It grows up to 3.1 m long on Guam (2.3 m in many parts of its native range), but the total length of most specimens encountered on Guam is less than 1.2 m. Juveniles constitute a greater proportion of the population on Guam than in the snake's native range, where juvenile mortality is likely to be much higher. This snake's ability to swallow exceptionally large prey items—up to 70% of the snake's body weight (Chiszar 1990)—suggests that its predatory capabilities are broad and therefore can affect prey species of a large range of sizes. Adult snakes can take birds the size of aga adults, but juvenile snakes might be restricted to eggs and small nestlings. Males attain greater sizes than females, so the largest males might be less frequently found in arboreal habitats.

Distribution

The species is extremely adaptable and occupies a wide range of tropical and subtropical environments among the hundreds of islands and continental locales within its native distribution. It has the most southern and eastern distribution of the 25 species of *Boiga*, which range from southwestern Asia through the Philippines and Indonesia, extending from Sulawesi in Indonesia through New Guinea and the Solomon Islands and along a narrow subxeric to humid coastal areas of northern and eastern Australia as far south as Sydney. In many parts of its range in Australia, the brown tree snake is the most-abundant arboreal snake and indeed the most-proficient arboreal predators of birds (Shine 1991a). It tolerates disturbed habitats, forages both on the ground and in trees, and is a common inhabitant of human dwellings, where it can play a role as a predator of rodents.

Active at night, it seeks refuge from high temperatures and bright light during daylight hours in the forest canopy, in dense vegetation, and even on the forest floor under logs and other structures, in limestone crevices, and in a multitude of sites made by humans, including buildings, vehicles, and shipping containers. The use of the latter daytime refuges places the brown tree snake in contact with cargo moving from Guam to other geographic areas. The activity of the snake is influenced by both humidity and temperature in the most-temperate parts of its range, but in moist tropical regions, it is active around the year and might have a nonseasonal reproductive cycle. Snakes are active year-round, but differences in detectability, frequency of power outages due to snakes, and frequency of snakebites exhibit seasonal components. On Guam, snake activity seems to be greatest at the transition from the dry to the wet season (May and June) and lowest during extremely dry seasons (Fritts 1988).

Diet

The brown tree snake has an extremely broad dietary range, including frogs, lizards, other snakes, birds, rodents, and bird eggs. On Guam, it is known to consume lizards, birds, eggs, insects, mammals, and human refuse, including picnic scraps, chicken bones, and even meat wrappers, although such scavenging might involve only snakes that are starving. Lizards constitute the most common prey of snakes on Guam and of juvenile and subadult snakes in the native range; larger snakes tend to take proportionally more birds and mammals than smaller snakes do (Greene 1989; Savidge 1987; Shine 1991a). On Guam, lizards predominate in analyses of stomach contents in all sizes of snakes (Rodda and others in press c; Savidge 1988); both terrestrial diurnal species and arboreal nocturnal species occur in the diet. That pattern suggests that snakes on Guam undergo an active search for sleeping prey and ambush foraging for active prey. The dietary flexibility of the brown tree snake contributes to its ability to remain abundant even after large segments of the prey base dropped precipitously. This characteristic explains why the brown tree snake has been able essentially to extirpate most of an island's avifauna and still remain relatively abundant.

Predatory Mechanisms

To find prey, the brown tree snake depends on its acute night vision when visual stimuli are present, but uses chemical cues (especially nonvolatile molecules) when deprived of visual cues (Chiszar and others 1988; Kardong and Smith 1991). Its abilities to use volatile cues and heat as predatory cues remain to be determined. When active, the snake moves slowly in the vegetation or on the ground and responds to both prey and potential threats that it detects visually. When closely approached by a human or illuminated by a strong light, it temporarily freezes and then flees within 10–60 seconds if the stimulus is not removed. There is evidence that the snake alters its foraging strategies, moving from forest canopy to the forest floor or cleared road sides, as arboreal lizard prey is depleted (Rodda and others in press a). Thus, in some situations when snakes are found on the ground, the number of snakes is inversely correlated with the abundance of lizards detected in the foliage, and the proportion of arboreal lizards in the snakes' stomachs is lower than that in snakes collected from trees or in habitats where arboreal lizards are more abundant. Whether snakes are able to assess prey availability on the basis of chemical cues or are merely foraging in habitats on the basis of proximal capture success and search images is unknown.

Field studies in the brown tree snake's native range strongly suggest that hatchling and juvenile survivorship is low because of the lack of prey small enough for it (Rodda and others in press c). Such reduced recruitment into the adult population thereby limits population densities of the snake in its native range. In contrast, oceanic islands (such as Guam and other Mariana islands) have higher numbers of native and introduced species of lizards that are small enough to be suitable as prey for juvenile snakes. These islands might therefore be able to support high populations of snakes if snakes are introduced. Also, small birds poorly equipped with antipredator defenses against snakes were prevalent on Guam and undoubtedly contributed to the buildup of snake populations there; indeed, the smallest and most-abundant birds on Guam disappeared first as a result of snake predation (Aguon 1983; Engbring and Fritts 1988; Pratt and others 1979). Malaita and the Solomon Islands are examples of the disparity of snake densities in native range versus oceanic islands. The same predators—mammals, birds, or reptiles—of the snake on Malaita are present on Guam. The average sighting rate for brown tree snakes on Malaita was 5% of the average for the same searchers on Guam (Rodda and others in press c). Shine (1991b) concluded that neither habitat structure nor predation was a limiting factor for other snakes; Rodda and others (in press c) conducted statistical analyses of faunal similarities that supported this conclusion.

In Australia and New Guinea, some antipredator defenses might partially protect birds from snake predation, but no specific research has been done on this. However, the lower populations of snakes in their native range (limited by availability of food for juveniles) than on Guam could be the primary factor in reducing risk of predation on bird populations. Armstrong and Pyke (1991) pointed to the lack of correlation between successful nesting efforts of a nectivorous bird in Australia and the peak of nectar resources. They concluded that nesting success was related to reduced activity of the brown tree snake during the coolest months.

Nesting failure during warmer months was related to egg and nestling predation and to a lesser extent to nest abandonment presumably because of attack by predators. In the warmer months, nestlings disappeared between dusk and dawn; the disappearance rate was halved during the coolest months, when the brown tree snake was least active (Armstrong and Pyke 1991). Bell (1966) observed similar patterns of nest predation, presumably by reptiles—with decreased nestling loss during cooler months and increased predation in warmer months—in an area inhabited by the brown tree snake.

Most studies of nest failure in birds underestimate the potential predatory impact of snakes. Snakes can engulf eggs and young and leave no trace of the predatory act; therefore losses may be higher. For example, Ricklefs (1969) recorded nearly half as many losses of yellow-headed blackbird eggs because of disappearance from nests for unknown reasons (42) as because of known predation (102) and almost equal numbers of disappearances of young for unknown reasons (22) and known predation (21). Snake (or bird) predation could have been a major factor in the disappearance of both eggs and young; if so, nest predation for this data set was more than 50% higher than reported by Ricklefs (1969).

EFFECTS OF THE PRESENCE OF THE BROWN TREE SNAKE ON GUAM

The brown tree snake was first reported in Guam in 1953, but reports of unidentified snakes in the vicinity of the military port in preceding years suggest that it arrived and became established in the late 1940s. The snake is likely to have arrived in military surplus equipment shipped to and through Guam for the Lend Lease Program in China, salvage and smelting, or return to the US mainland. Structural variation in body color and markings throughout the snake's native range provides a basis for establishing the source population as that on the Admiralty Islands (Rodda and others 1992), which coincides with the location of the large aggregation of Allied troops and materiel on Manus, the Pacific command headquarters before the capture of Guam in 1944.

A large volume of military traffic and cargo flowed from the Admiralties to Guam, and so too moved the snakes that colonized Guam. Within a decade after the species' arrival on Guam, large brown tree snakes were increasingly encountered by residents near the harbor and central Guam; by the 1960s, bird populations in the central part of the island began to disappear (Rodda and Fritts 1992; Savidge 1987). In Guam's environment of abundant lizards, birds, and introduced rodents, the snake thrived and attained large populations; by 1969, the species had spread throughout the island, even though the birds of northern Guam did not decline noticeably until more than 10 years later (Rodda and others 1992; Savidge 1987).

On the basis of patterns of the disappearance of Guam's avifauna and the conspicuousness of snakes, peaks in the eruption of the snake population moved progressively toward the north end of Guam until the 1980s, when most of the remnants of Guam's bird populations disappeared or were most severely depressed (Fritts 1988; Savidge 1987). Savidge demonstrated high predation rates using several species of birds in snake traps spaced in forest areas of Guam where native birds were declining. Snake populations declined by as much as 50% within 3 years after the primary elements of Guam's forest bird fauna disappeared from northern Guam (Rodda and others 1992). Predation on aga eggs by the brown tree snake has been observed (Campbell 1996b), and the loss of nestlings and adults has been inferred from disappearances from nests and territories studied by personnel of the Guam Department of Agriculture Division of Aquatic and Wildlife Resources (DAWR).

Since the arrival of the snake on Guam, most of Guam's indigenous forest vertebrates have been extirpated. Too-few baseline data are available to reveal unequivocally the degree to which the snake was responsible for those losses, but, although the exact sequence of extirpations and their causes are incompletely known, several lines of evidence point to the snake as the primary cause of the loss of many of Guam's birds and bats and several lizard species (Rodda and others in press a). Guam's 3 bat species (Wiles 1987) might have been prey for the snake, but data on 2 bats that disappeared before 1980 are lacking, and human hunting has obscured the cause of the decline of the remaining species (*Pteropus marianamus*) and its failure to recover. However, the snake appears to prey on young bats, inhibiting recruitment into an otherwise-protected population (Wiles 1987).

Evidence that the snake played a major role in the extirpation of the vertebrates of Guam (Rodda and others in press a) includes geographic patterns of bird and lizard losses concordant with population eruptions of the snake; predation of snakes on the vertebrates that declined, including lizards, birds, and mammals; lack of indications of substantial habitat loss, disease, pesticide contamination, or other causal factors; vulnerability of, and recorded impact on, both native and introduced species; unprecedentedly high populations of the snake coinciding with the disappearance of the avifauna; reductions in snake populations on northern Guam after the birds disappeared; and lack of comparable extirpations on nearby islands and islets where the snake is not present or conspicuous.

Estimates of snake density suggest unprecedented numbers of snakes on Guam (Rodda and others 1992, in press a). Snake densities in 1985, when the native forest birds of northern Guam disappeared or were reduced to only a few individuals, are estimated to have been about 100 per hectare (Rodda and others 1992). That estimate is based on trapping results and the calculated abundance of snakes in night surveys relative to results of work from 1987 through 1992, when snake populations were judged to have dropped to 35–50 snakes/ha—still 10 times greater than in its native range (Rodda 1996)—after the virtual collapse of most bird populations and the depression of rat, shrew, and some lizard populations. That the populations of the snake were exceptionally large and remain relatively large can be inferred from the frequency of power outages on Guam attributable to snakes' coming into contact with high-voltage conductors (Fritts and others 1987); the frequency with which snakes, potentially motivated by hunger, invaded homes and attacked sleeping humans (Fritts and others 1994); and the high incidence of snakes discovered in extralimital situations as a result of surface-ship and air traffic from Guam (Fritts 1987; Fritts and others in press).

Population estimates for snakes at individual locales have varied from 1988 through 1995; both increases and decreases have been noted, but the mean population estimates aggregate in the range of 35–58 snakes/ha. That the brown tree snake is not likely to exhaust its lizard prey base is suggested by 1995 studies of densities in forested areas (Campbell 1996a)—although there were 36/ha more, there were nearly 400 lizards and one rat for each snake. Thus, there is no scarcity of small prey, although the scarcity of moderate-size to large prey in many situations on Guam could place dietary limits on large snakes. The brown tree snake can survive (or even thrive) on a diet of small lizards and reach reproductive size (980 mm snout-to-vent length), thereby ensuring continuous recruitment into the breeding population.

Although large snakes can subsist on small prey, they might alter their foraging patterns in search of prey of more suitable size (by foraging over larger distances and using habitats where endotherms remain). Evidence suggests that snakes prolong foraging and movements into the early daytime hours: more electric outages have occurred in morning hours (6 AM to noon) since 1985 than occurred previously, when more than 95% of outages caused by snakes occurred at night (6 PM to 6 AM). The increase in morning activity without a comparable increase in afternoon activity suggests that snakes are prolonging nighttime activity rather than merely becoming more diurnal (Fritts and Chiszar in press). The dependence of large snakes (over 2 m long) on endothermic prey is evident in the scarcity of large snakes in forested areas and their prominence in areas around ranches and residential areas, where poultry, rodents, and introduced birds persist.

Thus, the decline of endotherms in native forests might have affected the size structure of the snake population without lessening the predatory pressure on the remaining endotherms, especially such forest inhabitants as the aga, the Mariana fruit bat, and the introduced black drongo and Philippine turtle dove. An incident in which a relatively small snake was discovered leaving an aga nest with an egg in its mouth suggests that three-quarters of male and more than half of female snakes are large enough to consume crow eggs and that the predatory pressure is high on nests in situations where snake densities are 30–50 snakes/ha. The density of snakes might have fallen more in natural habitats than in disturbed ones close to development: It is these disturbed habitats that produce snakes most likely to enter the transportation network, interrupt electric transmissions, and attack humans.

DECLINE OF BIRDS ON GUAM IN RELATION TO THE BROWN TREE SNAKE

Of the 22 native birds known to be resident on Guam, 13 have been extirpated (including 2 endemic species and 3 endemic subspecies) solely or primarily because of snake predation and related factors. Two endemic taxa (a rail and a kingfisher) have been extirpated from Guam but exist in captive situations. A few aga are also in captivity.

Four bird taxa are presumed to have disappeared before the arrival of the brown tree snake; at least, their absence cannot be directly linked to snake predation (Engbring and Fritts 1988). They include a single forest bird (megapode), which disappeared before the turn of the 20th century, and 3 oceanic or wetland birds (shearwater, Mariana mallard, and white-browed crake). Nine native forest birds apparently disappeared because of the snake. Another 3 oceanic birds (tern, noddy, and tropicbird) listed by Engbring and Fritts (1988) are biologically extirpated (infrequent sightings over the last decade involve birds arriving from Cocos or other islands); all formerly bred on Guam and no longer do so.

The 3 surviving native forest birds (aga, swiftlet, and starling) that remain on Guam occur in extremely reduced populations; of the 3 extant wetland birds that remain (bittern, reef heron, and moorhen), only the bittern appears common.

The aga is presumed to number between 20 and 40 in 1996. The swiftlet occurs as a regular inhabitant of a single cave, but recent estimates of swiftlet numbers suggest a decline from 3 years ago. An additional colony of fewer than 50 birds might exist periodically. A third colony is rumored to exist, but its size and location are unknown.

Assessing the impact of snake predation with indirect evidence on the 8 introduced species is more difficult because it is less evident what the densities and distribution of these species would be without snake predation. Over the last decade, all introduced species have been reduced, but some have periodically shown signs of population increase or succeeded in special situations, and that poses a dilemma in assessing their role as prey for the snake. Introduced species surviving on Guam could provide crucial information on methods of protecting native species, but of the introduced taxa only the black drongo (Maben, 1982) and Philippine turtle dove (Conry 1988) have been studied.

At present, introduced species are the only birds seen with any frequency on Guam, and even these are infrequent enough to be noteworthy for the lay observer. Four of the 8 introduced species of birds are reduced in numbers. The Eurasian tree sparrow is found primarily in urban, commercial, and roadside situations but has not successfully invaded forests. The black drongo and the Philippine turtle dove are conspicuous birds, but drongos are substantially reduced on the Northwest Field of Andersen Air Force Base in comparison with densities in 1985. Both drongos and turtle doves are present in low numbers over large areas of Guam, but they are obviously rarer than they would be if snake predation were not high (Conry 1988). Other introduced species are restricted in numbers and are found only in specialized situations affording partial protection from snake predation. For example, the francolin, mannikin, and quail are all typically grassland inhabitants, and the rock dove, which is an urban dweller, is extremely rare. Mannikins are infrequently encountered but only in grassy habitats. The status of the introduced coturnix quail is unknown; Engbring and Fritts (1988) considered it to be common. The status of feral populations of the jungle fowl is uncertain because of the prevalence of free-ranging chickens associated with ranch and suburban settlements, but, in direct contrast with other islands where wild populations of jungle fowl persist, it is likely that the jungle fowl is extinct on Guam except for recent escapees.

In summary, 4 native species are gone, and their disappearance cannot be attributed to the snake. However, 12 species probably have been extirpated by snake predation, and 6 are extant, but only one (the yellow bittern) is considered common. Snake predation has probably reduced populations of at least half the introduced species; grassland and urban species remain more abundant than forest dwellers. Of the introduced birds, only the Philippine turtle dove, black francolin, and Eurasian tree sparrow can be judged common.

It is perhaps notable that of 21 avian taxa endemic to the Marianas listed by Pratt and others (1979), 10 have disappeared from Guam and only 4 (nightingale reed warblers [*Acrocephalus luscinia nijoi* and *A. luscinia yamashinae*], collared kingfisher, and cardinal honeyeater) are now found north of Saipan. Another 11 taxa will be lost if the brown tree snake eliminates the forms now limited to the southern Northern Mariana Islands (Rota, Agiguan, Tinian, and Saipan). The threat of the snake's colonizing Saipan, Tinian, and Rota is related to increasing surface and air traffic between Guam and these islands. Agiguan is uninhabited and relatively inaccessible to humans, but its small size would increase the vulnerability of its bird fauna if the island were colonized by snakes.

EFFORTS TO CONTROL THE BROWN TREE SNAKE ON GUAM

The recognition that the brown tree snake was causing serious problems on Guam led to the development of means to control it. No easy solutions were apparent, nor was one agency or government solely responsible for leading or funding snake-control efforts. This section reviews the types of control methods and their effectiveness.

Trapping

Biological research by DAWR, the National Biological Service (NBS), and the US Department of Agriculture Animal Damage Control (ADC) led to the development of an effective snake-trapping protocol. Since its development, trapping has been the primary control strategy. DAWR biologists have trapped hundreds of snakes in the vicinity of endangered-bird habitats. The ADC program has trapped thousands of snakes annually (for example, 6,134 were trapped in 1995) primarily in or around transportation facilities (ports and airports). However, even a large number like the 1995 yield would theoretically account for the population of snakes from an area of only 123 ha (assuming 50 snakes/ha in forested habitats), but this same number applied to thousands of hectares represents only a small fraction of the snake population and is of negligible ecological benefit. Traps might be appropriate in small areas but are inadequate by themselves for larger areas because of manpower limitations and large populations and seasonal mobility of snakes.

Nest Protection with Barriers on Trees

Aguon and others (in press) have protected individual nests of aga by encircling the trunks of the nest trees with electric barriers (Figure 3-1) and pruning canopies of adjacent trees and vegetation to prevent snakes from crossing into the nest trees via canopies. Initiating control only after identification of an active nesting attempt can lead to human disturbance of nest-building, egg-laying, and incubation; however, DAWR staff report that aga have renested in trees that were protected by barriers the previous breeding season suggesting that the human disturbance of emplacing the barriers might not cause abandonment of the nest. Such a risk must be considered in decisions as to how much trimming is needed to separate canopies and the extent of active trapping to remove snakes that might already be present in the canopies.

Such protection may not be adequate in itself to allow recovery of the aga in the face of large populations of snakes. Despite increasing egg protection, fledging success has been poor—possibly not because barriers on trees failed but potentially because canopy separation was inadequate, especially during windy periods, or because snakes remained in the trees despite efforts to trap them. Conclusive data are lacking, and few hypotheses have been tested adequately, because there are so few nests. Isolation of crows from snakes must be increased to increase survival of crow eggs, nestlings, fledglings, young birds, and adults.

Snake Exclosures

The only adequate devices for controlling snake populations in natural areas are experimental snake exclosures (area barriers) constructed in a cooperative effort involving NBS and Ohio State University (Figure 3-2). An 18-month field study demonstrated the efficacy of a snake exclosure; snakes were removed from the exclosure by trapping and hand capture. Lizards, snakes, and birds were monitored in exclosures and in comparable sham plots from which snakes were not removed. Results detailed by Campbell (1996) suggested that exclosures facilitated removal of snakes from 2 1-ha plots (in 17 and 78 days) and allowed

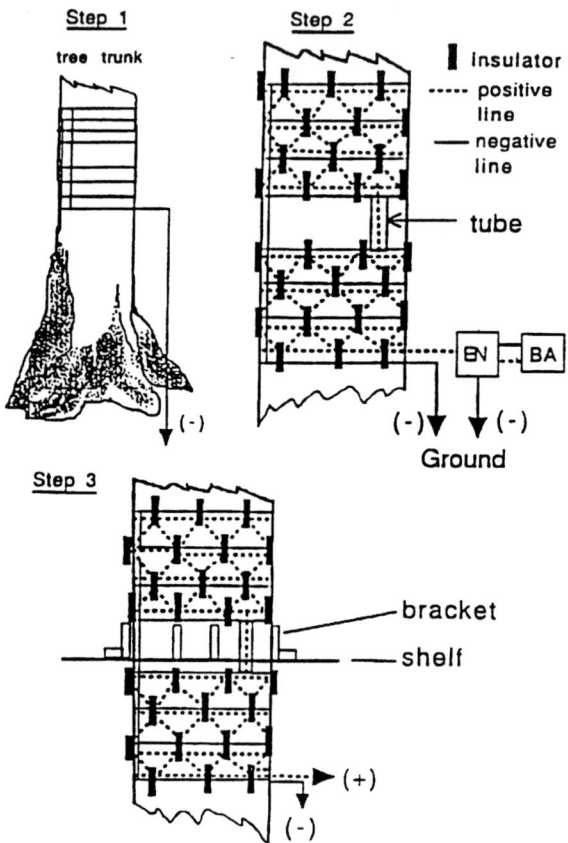

FIGURE 3-1 Electric barrier. 1) Galvanized cable wrapped around tree trunk. Steel rod inserted into ground forming negative line. 2) Insulators nailed to trunk in staggered pattern. Seperate length of cable placed on insulators forming positive line. 3) Positive line connected to electric fence energizer (EN) powered by 12v DC lead acid battery (BA). Metal shelf attached to tree trunk between upper and lower electrified sections. Source: Aguon and others, in press.

maintenance of effectively snake-free plots; about 5% of the number of snakes recorded in comparable habitats without controls were recorded in plots during the experiment, and a substantial increase in lizard abundance was evident in snake exclosures relative to that in plots with snakes. Problems of maintaining such exclosures were identified: damage by rats and large mammals, plant control, and the potential for typhoon damage. A second phase of exclosure design and testing with materials resistant to those problems was initiated in a laboratory setting where snake performance was monitored with an infrared-sensitive video

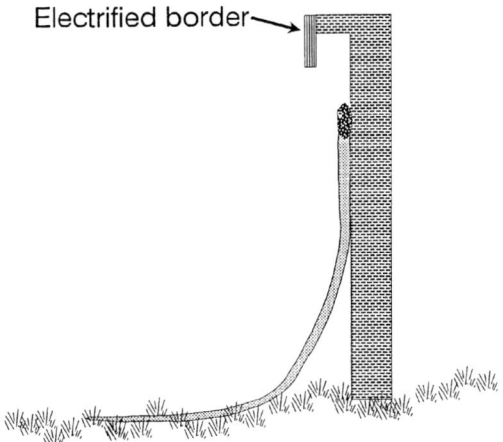

FIGURE 3-2 Snake exclosure barrier. Side view of concrete barrier wall. Source: committee generated.

recorder. New exclosure designs incorporating the following features were considered:

- Strength to resist or ability to shed wind loads during periods of high wind velocity;
- Surfaces smooth enough to discourage climbing; and
- Ledges, overhangs, or electric skirts to discourage overreaching (especially by large snakes).

One exclosure design, which uses masonry materials with a height of 1.15 m and an overhanging ledge fringed by an electrified metal strip 6 cm wide, has withstood over 1,000 attempts to enter by snakes of various sizes without being breached (Figure 3-2). Even without electrification of the 6-cm metal strip, 96% of snakes were unable to cross the same exclosure in hundreds of trials. An alternative design involving retrofitting of existing Cyclone fences is being tested and developed (Campbell 1996b).

Rodda and others (in press b) probed the critical elements of controlling snakes in small plots. Recent results suggest that snake exclosures (1–50 ha) applicable to conditions on Guam are feasible. Testing of larger-scale applications of exclosures under outdoor conditions is planned, and engineering-fabrication suppliers are involved in estimating costs of prototypes and applications.

As envisioned by Campbell (1996), snake exclosures have applications in a variety of contexts:

- to prevent snake from entering high-priority locations, such as endangered species habitats (to protect all species present), cargo storage or dispatch areas (to reduce export of snakes), electric substations and generating facilities (to reduce power outages), and residential neighborhoods and poultry farms (to protect humans and domesticated animals);
- to facilitate detection, capture, and containment of snakes arriving at extralimital sites in cargo (to interdict snakes) by constructing enclosure barriers that prevent snakes from leaving an area; and
- and to deflect movements of snakes toward traps or bait stations (to enhance control efficacy), especially in destinations judged to be at risk.

PROGNOSIS FOR CONTROL ON GUAM

The ability to control snake populations throughout Guam depends on the future development of the control tools identified in the integrated pest-management plan (BTSCC 1996; Campbell and others in press), which integrates large-scale control (such as biological or toxicologic methods) with other tools available now (trapping, hand capture, habitat management, repellents, fumigants, barriers, exclosures, and so on). Biological control and toxicants are not likely to be available within 3-5 years, and even then only if funding called for in the Aquatic Nuisance Species Task Force's Brown Tree Snake Control Plan (BTSCC 1996) is continued or expanded. However, the application of combinations of exclosures, barriers, traps, and other available techniques could be initiated in the near future to protect sites that have high priority for conserving endangered species. The size, location, and intensity of management of such sites should be determined carefully. Protection of areas, rather than solitary nest sites, has an added benefit of enhancing viability of several species, instead of a single taxon of special interest. In as much as eradication of snakes on Guam is distant or even unattainable, management of smaller areas offers the only in situ method for protection of endangered species. Exclosure barriers—whether physical, ecological, or geographic—will be important in reducing the required effort and increasing the efficacy of localized control programs.

THREAT OF SPREAD OF THE BROWN TREE SNAKE TO OTHER ISLANDS

In addition to Guam, the brown tree snake has been reported on 10 other islands—including islands in the Northern Marianas, the Federated States of Micronesia, the Ryukyu Islands, Diego Garcia Atoll, the state of Hawaii, and Wake Island—as a result of dispersal from Guam (Fritts and others in press). A single record exists of a brown tree snake's arrival in Corpus Christi, TX, in a shipment of military household goods from Guam (McCoid and others 1994). Such sightings involve snakes in civilian and military transportation systems, ships, and air traffic from Guam. Seven brown tree snakes have been verified in Hawaii (1981–1994); however, on the basis of captures of snakes and reports that did not result in captures, Saipan, Rota, and Tinian in CNMI are at greatest risk for infestation because of their proximity to Guam and the volume of cargo transported from and through Guam to these destinations. Since 1985, more than 30 sightings of snakes on Saipan, an island formerly lacking snakes, have been reported.

The likelihood of dispersal of the brown tree snakes as passive stowaways in ship and air traffic from Guam is increased by the central position of Guam in the transportation system of the western Pacific (especially for American flag islands). Increasing traffic volume is a direct result of commercial activity in the region. The incidence of brown tree snakes in cargo from Guam is also a result of the extremely high snake populations on Guam and the snakes' tolerance for disturbed habitats, including commercial, residential, and industrial sites from which cargo is shipped. The brown tree snake avoids high temperatures and bright sunlight by seeking daytime refugia. Vehicles, containers, construction materials, and other cargo furnish ample hiding spots and result in dispersal to and through ports and airports. Snakes can remain inactive and hidden and can survive without food for long periods. For example, the snake that crawled from household goods in Texas might have been in transit in a shipping container for up to 6 months.

As a native of hundreds of Pacific islands and a successful colonist of Guam, the brown tree snake is likely to succeed in establishing populations on other islands with similar environments if enough individuals arrive. As a habitat generalist and dietary opportunist, it might reach pest levels elsewhere in the Northern Mariana chain, elsewhere in Micronesia, in the state of Hawaii, and beyond unless dispersal from Guam is reduced, dispersing snakes are intercepted on arrival, and incipient populations are eradicated before they are widespread and firmly established. The environmental consequences on the Pacific islands that the snake has colonized might develop faster than on Guam because most of the islands are smaller than Guam. In many cases, the islands at risk hold some of the same—or at least closely related—bird and reptile taxa that were lost from Guam. Further losses would constitute a heightened threat to the biodiversity of the region. The long-term consequences of major disruption of entire bird, bat, and lizard faunas on Guam and other islands are poorly documented, but these organisms played ecological roles as insect predators, pollinators, seed dispersers, and consumers of fruit that are now only partially filled by introduced species and invertebrate organisms.

The brown tree snake constitutes a threat to the survival and conservation of the aga on two levels:

• predation on crows and interference with recruitment in the extremely reduced population on Guam, and

- risk of continued dispersal from Guam to other western Pacific islands, including Rota, where establishment and eruption of brown tree snakes would threaten the survival of the only other population of the aga.

Colonization of Rota would effectively nullify all management programs to protect and preserve the aga and its habitat. Approaches that will be necessary to address those two facets of the snake problem might well differ.

Efforts to Prevent Spread of the Brown Tree Snake to Other Islands

The documented dispersal of snakes from Guam to other islands and parts of the world (Fritts 1987; Fritts and others in press) resulted in the initiation of the ADC program to control snakes in the transportation system on Guam. Great strides have been made in and around ports and airports in implementing trapping, visual searches, canine inspections, and, to a lesser extent, management of habitats, but the enormity and complexity of the task remain evident. Recent discoveries (in November and December 1995) of snakes on ships and in cargo on Saipan and Pohnpei highlight the need to expand and enhance control efforts in the overall transportation network on Guam and on islands at risk of receiving snakes. Snake control focused on transportation from Guam is not complete: problems include limited ADC personnel on Guam, relying on voluntary compliance with ADC snake searches, periodic ship departures with no advance notice, and the complex dynamics of moving commercial, private, and military cargo in a heavily used port and airport system.

4

Major Findings and Criteria for Recovery of the Aga

To develop the scientific criteria for a recovery plan for the aga, the committee reviewed and analyzed the available scientific information, was briefed by biologists from the US Department of the Interior Fish and Wildlife Service (FWS) and National Biological Service (NBS), the Guam Department of Agriculture Division of Aquatic and Wildlife Resources (DAWR), and the Commonwealth of the Northern Mariana Islands (CNMI) Department of Lands and Natural Resources Division of Fish and Wildlife (DFW), and spoke with others with relevant expertise, including US Department of Agriculture Animal Damage Control (ADC) personnel. Based upon the information detailed in the preceding chapters, this chapter summarizes the committee's major findings on the current state of knowledge of the aga's plight, and enumerates the basic criteria for prioritizing our recommended recovery actions. Chapter 5 evaluates various recovery options and compares the aga's situation with that of other corvids, including the 'alala in Hawaii. Finally, chapter 6 offers 11 specific recommendations for recovery actions to be taken in the next three years.

SUMMARY OF MAJOR FINDINGS

1. The present population of the aga on Guam is not reproducing and is not viable. The species was once distributed throughout the island of Guam, but is now found only in limited forested areas of Andersen Air Force Base at the northern end of the island. The number of adult birds has declined from an estimated 50 in 1992–1993 to 20 in 1996. There are no known breeding pairs at present and only 7 nestlings are known to have fledged on Guam since 1986. Available

information strongly suggests that males outnumber females in the current population, and that the age structure of the population is skewed towards older birds.

2. The only viable population of the aga is on the island of Rota and it appears to have declined by at least 50% during the last decade. The most recent estimate of 592 aga on Rota in 1995 was made with a survey method (variable circular plots) that the committee believes is likely to overestimate birds like crows (Ralph and others 1993). Even so, comparisons using the same transects and the same detection methods lead to estimates of either 50% or 70% declines since 1982, depending on how the data are analyzed. These rates of decline are substantially higher than estimates of rates of loss of forested habitat during the same period.

3. Predation by the brown tree snake is the most likely primary cause of the decline of the aga on Guam. Coincident with the decline of the aga on Guam has been an explosion in the population of the introduced brown tree snake. This snake is also the most probable cause of the extirpation of several other native forest birds on Guam. The fact that both the decline of the aga and other forest birds on Rota have not been more severe, and that the brown tree snake has not been detected on Rota, support the hypothesis that the brown tree snake is the primary cause of decline of the aga on Guam. The only management tool that has been used to mitigate the effects of brown tree snakes on aga has been the placement of electric barriers on nest trees. Additional studies are needed to assess the effectiveness of canopy breaks in stopping migration into nest trees and the effectiveness of trapping snakes that are present in nest trees when the barriers are installed. Snakes have been successfully eradicated from exclosures as large as 1 ha and trapping might be effective at reducing snake numbers in areas up to 7 ha. Neither intensive trapping of snakes nor use of exclosures has been tested on a scale of tens of hectares in predominantly forested habitats, the scale that will probably be required to protect the aga.

4. Inadequate precautions are being taken to prevent the spread of the snake to other islands. Although the brown tree snake has not been found on Rota in natural habitats, individual brown tree snakes have been found on at least 10 other islands, including Oahu in Hawaii, the US mainland, and elsewhere; a population might already be established on Saipan. With many forested areas on Guam still having densities of over 30 snakes/ha and the urgent need to prevent snakes from being carried in air or ship cargo to other islands, control of the brown tree snake must be given the highest priority.

The ADC program has attempted to reduce the dispersal of snakes in air and ship traffic from Guam since 1993. This program depends to a great extent on visual searches with hand capture, trapping, and detector dog and handler teams and to a smaller extent on habitat and prey-base management near transportation facilities on Guam. The large numbers of snakes removed from within, around, and in moderate proximity to the transportation network make evident the risk of dispersing snakes to other islands and the need for additional snake control on Guam and other sites where they are likely to arrive. The number of snakes caught in 1995 is large (6,134), but verification of control efficacy, increased efficiency, and more comprehensive coverage of the transportation system are needed to adequately safeguard other geographic areas. Adequate justification exists to increase ongoing efforts in scope and intensity, and to simultaneously expand research to develop better, more effective technologies applicable to the large scale of the problem and the need for a multilayered safety net to prevent snake dispersal and colonization. The NBS program has emphasized the development of barriers to dispersal to augment other control techniques, arranged as exclosures to create snake-free areas. Effective control will probably require an integrated program that employs several techniques simultaneously. Neither DAWR nor DFW has aggressive management programs to control the brown tree snake.

5. Although based on limited data, estimates of genetic differences between the Guam and Rota populations indicate that differences do exist but are very small. Unique genetic diversity might reside in the Guam population (Tarr and Fleischer 1996), but any management decision based on genetic considerations should be considered in conjunction with the demographic goals of maximizing productivity and minimizing mortality rates (Lande 1988).

6. Aga nests fail at very high rates on Guam, compared with rates on Rota and rates in other corvids. Aga eggs or chicks in unprotected nests were almost always eaten by snakes or lost to other causes. Efforts by DAWR to install electric barriers around nest trees and to isolate nest trees from the adjacent canopy has increased clutch sizes and reduced losses of nestlings to brown tree snake predation, but have not increased the number of fledglings. Many nests are apparently abandoned either before eggs are laid or just afterwards, or the eggs are being eaten by brown tree snakes before researchers can detect them.

7. Eggs that have been taken from aga nests for artificial incubation have shown a high rate of developmental failure. Either many eggs are infertile or the embryos die in early developmental stages. That could be because the parent birds are getting too old to breed normally, because incubating birds are harassed by brown tree snakes, or because of other factors. The current state of knowledge is insufficient to distinguish between senescence and other causes of developmental failure.

8. Little is known about the demography, ecology, and behavior of the aga on either Rota or Guam. Field work on Rota has been limited by lack of perma-

nent staff supported either by DFW or by FWS. On Guam, adult mortality has not been directly studied, but territory turnover rates suggest that aga are disappearing more frequently than expected relative to other corvids. Because previous work has not involved studies of marked birds, either banded or tagged with radios, little is known about survivorship rates, the specific reactions of crows to the brown tree snake, why the birds on Guam seem unable to complete nest construction, and why the proportion of eggs that are viable is so low. Too few demographic data are available for viability analyses to be useful at this time. Studies of radio-tagged birds on both islands could provide information on causes of mortality and environmental stresses, and could yield needed demographic data. Comparisons between Rota and Guam could indicate similarities and differences between the two populations that could lead to insights about limiting factors. Such work should be done in conjunction with an assessment of the size and stability of the Rota population. Every effort must be made to ensure that Rota remains snake-free and that adequate habitat for the crow population is protected.

9. A captive population of 6 male and 6 female aga exists on Guam and the mainland. Of 10 adult crows taken from Rota by the Mariana Archipelago Rescue and Survey (MARS) program for captive breeding between 1993 and 1995, nine survive. Although 8 birds have been held as pairs (some with inadequate space for reproduction), only 1 pair produced 1 offspring. The current ten MARs birds are held in mainland zoo facilities. Two eggs produced in the wild on Guam in 1995–1996 hatched in captivity, and the resulting juveniles are being held in Guam.

10. A limited amount of aga habitat is protected on Guam and Rota. Habitat protection on Guam consists of about 1,600 forested hectares owned and managed by the government of Guam and 9,000 forested hectares recently incorporated into the Guam National Wildlife Refuge, mostly on Andersen Air Force Base. On Rota, most of the land is privately owned, and pressure for economic development is increasing. In an attempt to ensure that adequate forest habitat will be protected as other areas undergo development, a habitat-conservation planning process has begun on Rota. While the amount of habitat set aside on Rota and Guam could probably support a viable population of aga, the fact that the established wildlife conservation areas on Rota do not currently designate any areas known to be occupied by crows argues for protection of additional forest habitat. Forest habitat-protection and conservation-planning efforts on Rota should be accelerated.

11. Despite substantial efforts by many people, the overall level of cooperation and coordination among agencies and individuals involved in recovering the aga needs improvement. Data collected during management activities go

unanalyzed for years and have not been made accessible to appropriate groups or individuals. Off-island expertise has not been well integrated into recovery efforts. A coordinated recovery program is needed for the aga with a planned system of communication among participants and regular review by experts outside the program. Recovery of the aga is a complex issue that will require major activities and an organizational support system that is not now in place.

CRITERIA FOR PRIORITIZING AGA RECOVERY ACTIONS

In this section, the committee outlines its criteria for recovery of the aga. These criteria were used by the committee in evaluating the potential management options enumerated in chapter 5 and in formulating the specific recommendations contained in chapter 6. The committee believes that the following criteria are essential to develop a comprehensive recovery program for the aga.

1. Recovery of the Guam population is unlikely unless the brown tree snake can be controlled over multiple crow territories or tens to hundreds of hectares of contiguous forested habitat.

2. The most secure populations of an endangered species will serve as the source of individuals for long-term recovery. These populations tend to be the largest, the fastest growing, or the ones that are least affected by the factors that cause the species to decline. The core population of the aga is the one on Rota. Its security must have high priority.

3. A single population, no matter how large, is not immune to deterministic or chance events that can cause extinction. Each additional population decreases the likelihood that a single catastrophic event could extirpate a species. The aga is especially at risk because both Guam and Rota are subject to typhoons on the average of once every two years. The Guam population in its current, precarious state does not constitute a viable second population, so it cannot be considered an adequate hedge against the loss of the Rota population. This highlights the importance of restoring the Guam population, but it also suggests that the feasibility of establishing a population on another island should be considered.

4. Captive breeding of animals for release in the wild is slow, expensive, rarely results in successful reintroduction, and is not a good long-term conservation strategy (Snyder and others 1996). Captive propagation could be considered if the Rota population suffered a catastrophic decline, but captive breeding might divert attention away from needed field efforts.

5. Consideration of genetic variation within and between populations should be incorporated into population management actions whenever possible but not

to the detriment of demographic goals. To the extent possible, within-population genetic variability in the Guam population should be maintained, but not at the expense of increasing the growth in numbers and populations of wild crows.

6. Management actions for the aga should be guided by research designed to determine the causes of decline and reverse the factors that limit the wild population. A coordinated management and research effort that facilitates the collection of data is needed to compare the ecology of aga populations of Guam and Rota, and to identify the limiting factors within each population. For example, such an effort would help to determine whether senescence or some other causes are responsible for the low viability of eggs on Guam. To the extent possible, key questions should be answered by using principles of experimental design: having control groups, using replication, and randomly assigning experimental groups to treatments.

7. Recovery programs should be part of overall efforts to conserve biological diversity and restore ecosystems. Endangered species recovery programs can promote the conservation of many species simultaneously if they are based on restoration of ecosystems. Because the brown tree snake has disrupted the entire forest ecosystem of Guam, its control in aga habitat could benefit other vertebrates. For example, areas in which snake populations were reduced by means of exclosures could be used as release sites for other endangered species, such as the Guam rail or the Micronesian kingfisher.

5

Options for the Management of the Aga

This chapter describes various recovery options for the aga, and compares the advantages and disadvantages of these options.

CESSATION OF MANAGEMENT

If current management efforts were to cease on Guam, the aga would most likely be extirpated on the island when the current adults die. There has been no nonassisted production of young in the wild since the 1980s (Wiles and others 1995), and only 5 young have fledged since 1990 (Aguon 1996). In the absence of an effective snake-control program, it is unlikely that the native crow population in Guam can be sustained, much less increased. Although the Rota population is not being actively managed, the population there is decreasing. Furthermore, unless effective control of the brown tree snake and quarantine measures are developed and implemented, there is a high probability that snakes would eventually colonize Rota and drive the aga to extinction.

CONTINUATION OF PRESENT MANAGEMENT STRATEGY

Because of the unique circumstances posed by the brown tree snake on Guam, many management techniques developed for other endangered birds are inappropriate and cannot be applied to the aga. Although management efforts are evolving quickly, the current management procedures on Guam remain experimental.

Present management efforts regarding the Guam population have not resulted in either population stability or growth.

Reestablishment of a viable crow population on Guam is highly improbable unless more-effective snake-control procedures are developed and implemented on a larger scale in crow habitat. It must be noted, however, that the current experimental work on snake control being carried out by the Guam Department of Agriculture Division of Aquatic and Wildlife Resources (DAWR) and US Department of the Interior National Biological Service (NBS) personnel might become essential for developing better crow-management strategies on Rota or in places into which the crow might be introduced (see "Translocation Options" later in this chapter). Although there is inherent benefit in continuing this work, it is not likely to result in the recovery of the Guam population. On Rota, the persistence time of the crow population could probably be increased through a combination of protection, public education, and habitat preservation, unless the brown tree snake colonizes the island. Without effective snake inspection, cargo quarantine, and control methods, such colonization is likely to occur and compromise the Rota population.

INTENSIVE MANAGEMENT

Intensive management involves the direct manipulation of nesting pairs in an attempt to increase the reproductive output of the population. The typical strategy is to remove eggs from wild pairs, artificially incubate them, hand-rear the resulting young, and either foster the young back into wild nests or release them with a soft-release technique (NRC 1992). Those methods have worked well on a variety of species, including European corvids (Delvaux 1991), North American corvids (Marzluff 1993; Marzluff and others 1994), and the 'alala (Hawaiian crow) (Derrickson 1994; Kuehler and others 1995).

The advantages and disadvantages of intensive management are discussed in a National Research Council report (see table 6.1 in NRC 1992), and they remain relevant for the aga with the following modifications based on recent experience with common ravens (*Corvus corax*), American crows (*Corvus brachyrhynchos*), black-billed magpies (*Pica pica*), and Hawaiian crows (Derrickson 1994; Kuehler and others 1995; Marzluff 1993; Marzluff and others 1994).

Intensive efforts to recover the 'alala began in 1992 after the Research Council committee's report (NRC 1992). Much has now been learned that is directly applicable to recovery of the aga. In particular, the technical aspects of observing, monitoring, and managing the wild population of 'alala are directly transferable to needed field studies and nest manipulations on Guam and Rota. Captive-propagation techniques developed for the 'alala are also directly applicable to the aga with slight modification, as discussed later in this chapter.

Removal of Eggs from Wild Nests for Artificial Incubation

Advantages of egg removal are more certain now than they were before the 'alala-restoration project. Hatchability and survivability of eggs and chicks are increased in captivity, and the annual cohort of young is much larger with intensive management than without it (Derrickson 1994; Marzluff 1993; Marzluff and others 1994). Most of the problems previously encountered with this technique have been overcome. Optimal techniques for artificial incubation have been developed and already applied to aga eggs (Brock 1996). A key is to allow parents to incubate eggs for 5–7 days before egg removal. The concern that all nesting pairs will not lay another clutch is also diminished. Corvids almost always renest after clutch removal (Marzluff 1993; Marzluff and others 1994); given the propensity of aga to renest after natural failure (Morton 1996a), we expect them to renest readily after clutches are removed. Hatching and rearing chicks in captivity have the disadvantage of requiring expanded staff and facilities, as in the case of the 'alala.

Soft Methods of Release

Soft methods of release range from controlled release over a period of days or weeks (hacking) to more elaborate systems involving holding birds for extended periods in outdoor flight cages in the habitat where they are likely to be released. Disadvantages associated with this technique (NRC 1992) are greatly reduced by the knowledge gained during 'alala restoration. The technique remains labor-intensive and expensive, but there is now less concern that captive-reared birds will not learn appropriate survival or social skills. Captive-reared corvids readily integrate with wild birds after release and have no apparent problems in socializing, foraging, or avoiding predators (Banko 1996; Marzluff 1993; Marzluff and others 1994). If birds are to be held for long periods (over 6 months), wild birds can be housed with captive-reared birds as tutors to facilitate social integration (Marzluff and others 1994).

A key to successful release of corvids is to raise them in groups and allow them to acclimate to the release area in a flight cage for several weeks or months before release. If corvids are raised with other conspecifics from hatch through release, puppet-rearing and tutoring are not necessary to enhance survival or breeding. Captive-reared black-billed magpies and American crows raised without puppets or tutors are breeding successfully in the wild.

The goals of a release project influence the type of rearing used. If the goal is to produce corvids that will readily assimilate with wild birds, tutoring and releasing birds when they are several months old can be beneficial (Marzluff and others 1994). However, if the goal is to colonize a new area and managers want released birds to remain near the release site, techniques for improving socialization and rapid acquisition of independence, such as tutoring, puppet-rearing, and maximizing time before release, might actually be detrimental (Marzluff and others 1994).

The advantages of soft releases listed in the Research Council report (NRC 1992) are all relevant to aga. The ability to colonize vacant portions of the range is especially beneficial. In fact, releases in vacant areas are the most likely to establish new breeders in the population quickly, apparently because territorial defense in occupied habitats reduces the ability of released birds to establish their own territories. The only hand-reared corvids known to breed in the wild were released in previously vacant areas.

Fostering as a Method of Release

Fostering nestlings to wild pairs is the least expensive and most natural way to release birds born in captivity. It will probably be an important option for aga when reproductive success in the wild returns to normal, and it remains a good option when only 1 chick is raised in captivity. Results of experiments with American crows, black-billed magpies, and common ravens suggest that chicks will be readily adopted even when several weeks old (Marzluff and others 1994). Fostering older chicks (10–20 days old) is preferred because they are less vulnerable to parental neglect or aggression. Older chicks can be substituted for newly laid eggs without apparent problems (Marzluff and others 1994).

Monitoring of Released Birds

Released birds should be closely monitored so that knowledge about limiting factors, sociability, and productivity can be gained. Radiotelemetry techniques have been developed for 'alala and tested successfully on aga, and 1 radio-tagged aga has nested in the wild (Morton 1996a).

TRANSLOCATION OPTIONS

Conservation programs that involve translocation (moving animals between populations, moving animals into unoccupied former habitat, or releasing animals into new habitat) are biologically risky, expensive, time-consuming, and usually unsuccessful (Ebenhard 1988; Griffith and others 1989; Snyder and others 1996). Translocation should never be attempted except as a last resort, and even then only after comprehensive planning to ensure the long-term survival of both the donor and recipient populations and their communities. Various potential translocation options for the aga are discussed below with recognition of the concerns raised above regarding the success of past programs for other species.

> ### The Impracticality of Translocating Birds from Rota to Guam
>
> In 1995, DAWR submitted to FWS a request for a permit to translocate nestlings from Rota to Guam. The committee used a Leslie-Lefkovitch matrix model to estimate what level of reproductive success would be needed to bolster the Guam population. If annual survivorship from fledging to adult were 0.74, then fecundity would need to be 0.5, or about 20 times greater than it currently is, for the population to sustain itself. The annual production of two fledglings per territorial pair (not per nesting pair) would be required for lambda to equal one (for the population to sustain itself). Inasmuch as one-quarter of the chicks in protected crow nests on Guam died before fledging (Aguon 1996), sustaining the population would require 2.67 young to be fostered per territorial pair [2/(1–.25)] to fledge two young per territorial pair. If there were eight territorial pairs, as at the beginning of the 1995–96 breeding season, about 21 chicks would have to be placed into crow nests for the population to sustain itself. This number would necessarily increase as more territorial pairs were reestablished. Although these predictions are very rough and are subject to the caveats stated in chapter 2, they are consistent with the committee's conclusion that the introduction of eggs or young birds from Rota would not be a practical solution to the problem of the decline of the Guam population with the current rates of adult survival.

Translocation from Rota to Guam

There are many more crows on Rota than on Guam, so an argument could be made to translocate some birds to Guam as a population enhancement strategy. Such a translocation would presumably serve 2 immediate objectives: the Guam population would be increased, and reproductive problems related to senescence would be rectified. However, with the current tree-barrier method, even a relatively small increase in the Guam population will require substantial increases in time, personnel, and resources. It is very unlikely that a large increase in the Guam population will be possible until snakes are controlled in or excluded from extensive areas of suitable habitat. Although removal of eggs or nestlings from the Rota population would probably entail little social or demographic disruption, given the renesting potential of this species, such removal cannot be justified biologically until the size and demography of the Rota population are accurately determined. Removal of birds from Rota might also jeopardize the present, politically sensitive habitat-preservation efforts on that island and hence the long-term security of its population.

Translocation from Guam to Rota

Because snake problems on Guam are so severe, it might be advisable to move some or all of the Guam crows to Rota. Moving some of the crows to Rota might increase the genetic diversity of the Rota population, preserve the genetic diversity of the Guam population by incorporating it into the Rota population, increase the demographic stability of the Rota population, and provide a greater chance of continued survival and reproduction of individuals from Guam. Problems to be dealt with include being comfortable with genetic mixing of the populations. Tarr and Fleischer (1995) have noted that the genetic differences between populations are not great enough to raise concern and have suggested that the demographic contribution might be more critical than genetic considerations at this point.

However, given the advanced age of the majority of the Guam birds, their demographic contribution to the Rota population would be minimal. Furthermore, taking birds from Guam would weaken the Guam population. Therefore, this option should be considered only after population-enhancement efforts on Guam are deemed unsuccessful. In addition, habitats would need to be carefully assessed because the aga uses somewhat different habitats on the two islands. Finally, both habitat availability and the social structure of the population on Rota would need to be characterized to ensure that suitable vacant habitat is available and that translocating birds into the population would cause minimal social disruption.

Removing all crows from Guam and putting them all on Rota would be a major event in the species's history and should be undertaken only after careful consideration. The translocated birds might have a better chance of surviving on Rota, but the extirpation of the Guam population would have far-reaching effects. First, having all aga on 1 small island is inherently unsafe and would substantially increase the risk of extinction of the species as a whole because of typhoons, disease, predators, or progressive habitat fragmentation and modification. Second, loss of another native forest bird from Guam could make further habitat-restoration efforts and snake-eradication research more difficult to defend. Third, loss of a native species from Guam would have to be discussed among many local groups, especially Chamorro groups in whose culture the species has been inherent for centuries. These drawbacks could be mitigated if, after brown tree snake control is substantially improved, birds from Rota were reintroduced to Guam.

Translocation from Guam, Rota, or Both to Other Islands

In most cases, the viability of severely declining species is improved when populations are established in several locations. Maintenance of multiple populations decreases the chance that one catastrophic event would cause extinction. Furthermore, genetic diversity might be better maintained because each population differentially experiences the effects of genetic drift, so fewer unique alleles are lost to the species overall (although each population will lose unique alleles). Finally, if each population expands, there will be greater demographic stability for the species.

Because the aga has occurred only on Guam and Rota during historical times, any effort to establish additional populations will entail releasing birds outside their historical range. Such introductions are generally not successful (Griffith and others 1989). Translocation of the aga to other islands would have to be evaluated carefully from the viewpoint of its contribution to persistence of the species and its effects on the flora and fauna of the other islands. Of course, the number, sex, social status, age, and so forth of the crows to be introduced would have to be considered.

The most likely places to search for new sites would be nearby islands where the habitat would be appropriate and the crow's presence would have minimal effects on other species. Some of the more-northerly, uninhabited islands or even the small island of Agiguan, between Rota and Tinian (Figure 1-1), might be appropriate; however, investigations to determine potential environmental impacts would have to be conducted before a selection is made. Other possible sites include islands in Palau, Pohnpei, or the Marshall Islands, where forest-bird assemblages are similar to those in the Mariana archipelago.

Movement of all of Guam's birds to a new site would have to be considered in light of potential genetic consequences. Because the Guam population appears to have greater genetic diversity than the Rota population, it might be advantageous to introduce some of Guam's birds to a new island and others to Rota to increase genetic diversity in an established population that is likely to persist at least in the short term. The same would hold true if only some of Guam's birds were to be translocated.

Translocation of some Rota birds to a new site should be undertaken with caution because it is the only successfully reproducing population of aga. However, if a complete and accurate census of the Rota crows indicates that there are unpaired birds, birds without nest sites, or birds whose territories are slated for destruction by development or that eggs can be harvested without adversely affecting productivity, there is probably good reason to invest in a third population.

Translocation from Captivity to Guam, Rota, or Other Islands

Releasing the captive birds back into the wild warrants serious consideration. Although the captive birds potentially constitute a third population themselves, they have not yet reproduced well and therefore cannot be considered a viable population at this point. The captive program was initiated for research and training purposes and not specifically to establish a viable captive population. Because husbandry techniques have been established for corvids in general, there is less need for a captive population. If a captive population were required in the future, one could be established on an emergency basis. It might be better to put the captive birds back into the wild if there were places where they could survive and contribute to population recovery or if monitoring their behavior would provide important information on their biology. This option keeps the focus of the program on research on and recovery of wild populations. Because most of the captive birds were wild-caught, they should demonstrate normal survival on release into the wild.

Given that all the captive birds in the Mariana Archipelago Rescue and Survey program were derived from the Rota population, their return to Rota would have minimal beneficial effects on the wild population's demographic or genetic structure. Furthermore, the risk of introducing diseases acquired in captivity to the population on Rota must be weighed carefully. Even if the risk is very low, the consequences would be so serious with respect to species survival that this option should be avoided. In contrast, given the precarious state of the Guam population and the low probability of short-term survival, the risks to the aga associated with introducing the captive birds into the Guam population are much more acceptable. Even if translocating the captive birds to the Guam population would not increase population viability substantially in the short term, it should allow for the refinement of practical translocation methods and the continuation of critical research on snake management and control, habitat use and preference, and the role of senescence in reduced productivity.

Translocation from Guam to Captivity

The Guam birds could be moved into captivity until effective snake-control methods are available. However, this plan presents numerous problems. First, it would constitute extirpation of the Guam population. Second, husbandry techniques have largely been worked out, so there is no purpose in keeping birds in captivity unnecessarily. Third, removing birds to captivity would preclude conducting critical field research. Fourth, removing birds from the wild could subvert current efforts to control the brown tree snake and to preserve wild populations and habitats. Fifth, captive breeding is a conservation technique with a number of important, but often-overlooked limitations, including difficulties in establishing self-sustaining captive populations, poor success in reintroducing captive-bred stock to the wild, progressive domestication, disease outbreaks, preemption of alternative conservation methods, high program costs, and difficulties in maintaining administrative continuity (Snyder and others 1996). Finally, establishing a viable captive population by using Guam birds alone would be impractical, given the age of most of the birds and the apparent age-related reproductive problems. Even if those problems could be overcome, offspring of the captive population would have to be released into new areas unless effective, large-scale methods of snake control or eradication were developed and implemented on Guam.

6

Eleven Recommended Management Actions for Recovery and Their Rationale

After careful review of the findings about the aga and the brown tree snake and consideration of the recovery options for the aga in the context of other endangered species programs, including those for corvids, the committee offers 11 recommendations for activities to take place in the next 3 years. The first 6 recommendations involve management and research; the remaining 5 involve infrastructure and support. Table 6-1 summarizes the proposed schedule. The committee does not recommend the continued maintenance of a captive breeding population or the transfer of crows from Rota to Guam during this period. The latter option should be considered only after studies indicate that translocation will not jeopardize the Rota population and that snake-control measures on Guam are sufficient to predict higher than current levels of survivorship of aga on Guam. Given the fact that two aga populations are extant at present, the committee strongly recommends that comparative, experimental studies be conducted.

RECOMMENDATION 1

Expand and increase research, development, and implementation of methods to control the brown tree snake and to prevent its spread to other islands. The agencies most in need of allocating additional funding are the US Department of the Interior (DOI) National Biological Service (NBS), the Department of Defense (DOD), the Guam Department of Agriculture Division of Aquatic and Wildlife Resources (DAWR), and the Commonwealth of the Northern Mariana Islands (CNMI) Department of Lands and Natural Resources Division of Fish and Wildlife (DFW). Recovery of the Guam aga population will not be possible unless the

TABLE 6-1 Simplified Outline of Recommended Management Actions for Aga Recovery

Management Activities	Year 1	Year 2	Year 3
SNAKE-CONTROL MEASURES			
Increase security at ports on Guam and Rota			
Plan cargo and quarantine areas	———————————————→		
Increase inspections by dogs on Guam	———————————————→		
Initiate inspections by dogs in CNMI, especially Rota.	———————————————→		
Accelerate research on snake ecology	———————————————→		
Accelerate research on integrated control methods	———————————————→		
Construct and emplace tree barriers on Guam	———————————————→		
Expand trapping effort in and around nest trees on Guam	———————————————→		
Look for direct evidence of snakes on Rota			
Trapping	———————————————→		
Night surveys	———————————————→		
If snakes found on Rota, use all available methods to eradicate or control	———————————————→		
GUAM			
Conduct annual censuses	———————————————→		
Intensive management of wild birds			
Collect and hatch eggs from wild pairs	———————————————→		
Band, radio-mark and release hand-reared birds (~6 months of age)	———————————————→		
Release both juvenile crows from DAWR facility with radios and monitor	————————→		
Release MARS birds and capture wild birds for marking using MARS birds as attractants	————→		
	5 females & 3 males	2 males	
Conduct telemetry studies (all birds, all years)	———————————————→		
Collect ecological information for comparison with Rota population	———————————————→		
ROTA			
Collect demographic and ecological information			
Survey and map all territorial pairs	———————————————→		
Conduct nest studies and obeservations	———————————————→		
Band nestlings	———————————————→		
Radio-mark fledglings and adults	———————————————→		
Investigate sources of mortality	———————————————→		
Develop and implement a habitat-conservation plan	———————————————→		
Make comparisons with Guam population.	———————————————→		
INTER-ISLAND TRANSLOCATIONS			
Evaluate potential translocation sites outside historic range	——→		
Conduct field examinations of potential translocation sites		——→	
RECOVERY TEAM			
Establish recovery team	——→		
Set and direct recovery program priorities	———————————————→		
Evaluate field efforts and meet every 6 months	———————————————→		
RESEARCH SCIENTIST/PROGRAM LEADER-COORDINATOR			
Create position and hire	——→		
Initiate program and coordination with recovery team	———————————————→		
Establish and maintain research databases and record systems	———————————————→		
PUBLIC EDUCATION PROGRAM			
Evaluate previous/current efforts	——→		
Establish objectives and target audiences	———————→		
Develop and implement program	———————————→		

effects of the brown tree snake can be mitigated. Furthermore, the most serious threat to the aga as a species is the spread of the brown tree snake to the island of Rota. The committee strongly recommends that a comprehensive approach to snake control be undertaken. It should include the following actions:

- Institute aggressive interdiction measures for snakes on Guam, such as exclosures in and around air and naval ports, to protect cargo going to Rota. The current US Department of Agriculture (USDA) Animal Damage Control (ADC) program is insufficient for this task. Have scientists and managers outside the agency conduct a critical review of the current ADC program as soon as possible. See chapter 3, "Efforts to Control the Snake."
- Conduct early detection and eradication programs for brown tree snakes on other islands, including Rota. Protect ports on other islands with cargo quarantine enclosures, fumigation, and inspection of high-risk cargo. If snakes are discovered in port or cargo-storage areas, deploy barriers to slow their dispersal from the initial sites of infestation. Ascertain the extent of infestation (density and area occupied) and conduct intensive eradication effort. See chapter 3, "Threat of Spread of the Brown Tree Snake to Other Islands."
- Establish a detector dog and handler program within the CNMI (especially Rota) to facilitate interdiction of snakes in high-risk cargo and other equipment that is not readily amenable to visual inspection. Test the effectiveness of the dog and handler teams. Use artificial training aids to solve problems related to training and maintenance of the dogs on islands where snakes are not available for training purposes. See chapter 3, "Efforts to Control" and "Threat of Spread of the Brown Tree Snake to Other Islands."
- Continue and expand research to improve brown tree snake control methods. Train all personnel involved in control efforts. See chapter 3, "Threat of Spread of the Brown Tree Snake to Other Islands" and "Prognosis for Control on Guam."

RECOMMENDATION 2

Study the behavior, population biology, and health of marked birds on both Guam and Rota. Identify limiting factors and collect data to elucidate the demography of each population. Make comparisons between the two populations with regard to fecundity, the causes and timing of adult and juvenile deaths, changes in home-range sizes, habitat use, foraging and roosting behavior, and, on Guam, responses to the presence and absence of brown tree snakes. This project will involve the following tasks.

- Capture and band as many young and adult aga as possible on both islands. See "Possible Causes of Population Declines" and "Past and Current Conservation Management" in chapter 2.

- Equip a subset of captured birds with radiotransmitters so that both diurnal and nocturnal monitoring can be conducted. See "Possible Causes of Population Declines," "Past and Current Conservation Management" in chapter 2, and "Monitoring of Released Birds" in chapter 5.
- At the time of banding, conduct complete physical examinations and health screens and collect samples of blood for genetic analyses. Using this information, obtain baseline data on the genetic variation in the populations, the prevalence of diseases and parasites, and normal hematology and clinical-chemistry values. See "Possible Causes of Population Declines" and "Past and Current Conservation Management" in chapter 2.
- Gather comparative information on behavior, ecology and movements on Guam and Rota. See "Possible Causes of Population Declines" and "Past and Current Conservation Management" in chapter 2.

RECOMMENDATION 3

Release all captive female aga and 3 of the captive males from the Mariana Rescue and Survey (MARS) program on Guam before the 1996–1997 breeding season. Pairs that formed strong pairbonds in captivity should be released together. Release unpaired females in territories occupied by suspected unpaired males. Release the captive birds from snakeproof aviaries equipped with funnel and noose traps built into them. Wild aga lured into the aviary should be captured and banded. Release the remaining 2 captive MARS males before the 1997–1998 breeding season. Release the 2 captive-reared juvenile aga now held on Guam before the 1996–1997 breeding season. Conduct thorough health screening of the captive birds before releasing them to the wild. All released birds should be radio-tagged and their behavior and interactions with snakes monitored closely day and night. See "Monitoring of Released Birds" and "Translocation from Captivity to Guam, Rota, or Other Islands" in chapter 5.

RECOMMENDATION 4

Place electric barriers on all nest trees to protect nests from brown tree snakes and monitor lizards, and increase trapping efforts in nest trees and adjacent areas. Test the effectiveness of nest-tree barriers, exclosure barriers, and trapping of brown tree snakes in increasing the rate of construction of nests and the production of viable eggs by aga. To do so, place tree barriers, exclosures, and traps for brown tree snakes in aga territories. See "Prognosis for Control on Guam" in chapter 3.

Collect all wild aga eggs after several days of incubation for intensive avicultural management. Allow some natural incubation of eggs by the female. Raise hatchlings in captivity, releasing them as juveniles on the fringes of occupied crow habitat before the next breeding season. See "Natural History" in chapter 2, and "Removal of Eggs from Wild Nests for Incubation" and "Soft Methods of Release" in chapter 5.

RECOMMENDATION 5

Assess the feasibility of translocating birds to islands other than Guam or Rota, including islands outside the Mariana archipelago. The committee favors preservation of the Guam and Rota populations over establishment of a population elsewhere. However, if the Guam population cannot be recovered, establish a second population on another island. If the recommendations outlined in this report are followed, data should be available within the next 3 years to determine whether translocation to islands outside the aga's historic range are warranted and desirable. However, before translocation to another island is considered, the following actions should be taken.

- Complete a survey of the flora and fauna of the island, with emphasis on endemic species. See "Translocation from Guam, Rota, or Both to Other Islands" in chapter 5.
- Assess the likelihood of establishing a self-sustaining population of crows on the island. See "Translocation from Guam, Rota, or Both to Other Islands" in chapter 5.
- Evaluate the potential negative impact of the crows on the existing flora and fauna, particularly with regard to endemic species. See "Translocation from Guam, Rota, or Both to Other Islands" in chapter 5.
- Assess the feasibility of eradicating nonnative species that might adversely effect aga. See "Translocation from Guam, Rota, or Both to Other Islands" in chapter 5.
- Consider the legal status of the crow on the receiving island. See "Translocation from Guam, Rota, or Both to Other Islands" in chapter 5.
- Conduct surveys to determine whether brown tree snakes are on the receiving island. See "Translocation from Guam, Rota, or Both to Other Islands" in chapter 5.

RECOMMENDATION 6

Conduct annual censuses of the aga population on both islands about a month before the breeding season. Conduct a complete census of the population and map all territories. Teams of observers should coordinate their efforts to count crows simultaneously in subregions (drainages) of the island. Appraise the feasibility of an annual census on Rota. If such a census would detract from obtaining estimates of productivity and survivorship, a random sample of territories could be monitored annually. Playbacks of recorded aga vocalizations might be useful for locating pairs. See "Past and Current Management: Guam" and "Past and Current Managment: Rota" in chapter 2.

RECOMMENDATION 7

Conduct complete postmortem examinations on all recoverable eggs and birds. A veterinary pathologist (preferably certified by the American College of Veterinary Pathologists) with experience in avian pathology should be enlisted for histopathologic examinations. From postmortem investigations, obtain accurate determinations of causes of death and information on endemic infections, parasitism, subclinical diseases, and nutritional condition. In addition, develop a tissue archive that can aid in future multidisciplinary research endeavors. See "Disease" in chapter 2.

RECOMMENDATION 8

As soon as possirle, develop and implement a habitat conservation plan for Rota that preserves essential crow habitat but allows economic development in an ecologically responsible manner. See "Past and Current Conservation Management: Rota" in chapter 2.

RECOMMENDATION 9

Establish a professional-level research-manager position on Guam within the NBS. The position should be filled only after a wide search for an appropriate research scientist. The person should be capable of leading a rigorous research and management program for the aga and constructing a database of information on it and other endangered birds in the Mariana archipelago. See "Past and Current Conservation and Management" in chapter 2 and "Finding 11" in chapter 4. Responsibilities for the research leader include the following.

- Ensuring effective communication among all recovery participants—DFW, DAWR, the DOI FWS and NBS, the USDA ADC program, DoD, conservation groups, and private individuals. Data should be kept in a centralized database and disseminated to all parties. The most effective and most efficient means of communication should be used, including e-mail, fax, electronic-database networks, and telephone conference calls.
- Ensuring that recovery efforts for the aga are integrated and coordinated with other efforts to manage and conserve the flora and fauna of the Mariana Islands.

RECOMMENDATION 10

Within the next year, appoint a recovery team specifically for the aga. The recovery team should include a crow biologist, an 'alala specialist, a population biologist, an avian veterinarian, a snake biologist, and an animal-damage control specialist. If possible, members should represent all agencies and organizations with vested interests. The recommendations contained in this report are designed primarily to guide the recovery effort only for the next 3 years. A recovery plan should then be developed on the basis of an evaluation of the outcome of the recommended actions. To that end, the recovery team should receive tabulated data and regular reports from the field. The team should meet regularly, monitor progress of recovery actions, and identify research and management priorities. The new program must include all stakeholders, and it must facilitate transfer of information among them. Establishment of a recovery team for the aga would facilitate cooperation among agencies and individuals involved in the recovery effort. It should include regular participation by off-island experts, and receive and respond to periodic review by experts not associated with the program. The team must be responsive to new developments and not get bogged down in writing a long, comprehensive recovery plan that will be useful for guiding efforts for only a very short time. Recovery efforts for the aga need to be coordinated with recovery efforts aimed at other species, but crow recovery is best served by a single-species team. The FWS Pacific Island Recovery Team is responsible for the recovery of a variety of endangered and threatened species in the Pacific Basin. It cannot possibly provide the management and research focus that the aga deserves. See "Past and Current Conservation and Management" in chapter 2 and "Finding 11" in chapter 4.

RECOMMENDATION 11

Conduct a public education program that specifically addresses issues relevant to aga conservation, including the problems associated with the spread of the brown tree snake. All of Micronesia and Hawaii should be targeted for education about the brown tree snake. See "Past and Current Conservation and Management" in chapter 2. The education program should

- Clearly define the problem, which might be different for different groups and different islands.
- Identify and specifically target all appropriate constituencies (such as schoolchildren, farmers, hunters, developers, policy-makers, and church and community groups) with appropriate methods.
- Focus feelings of national pride on indigenous wildlife, including the aga.
- Involve local and indigenous people in implementing the program.

- Incorporate materials previously produced by DAWR, FWS NBS, and ADC. These materials should be evaluated, expanded, and regularly updated to form a comprehensive educational program.
- Develop systems for monitoring, evaluating, and revising the program. Programmatic evaluations should be based on both internal review by staff and input from external target audiences and participants.

7

References

Aguon, CF. 1983. Survey and inventory of native land birds on Guam. Mangilao (Guam): Department of Agriculture. Division of Aquatic and Wildlife Resources Annual Report. p 143–57.

Aguon, CF. 1993. Development of testing of control methods for the brown tree snake. Guam Department of Agriculture, Progress Report—Federal Aid to Fish and Wildlife, Project No. E-1-8, Study No. 7, Job No. 3, 3 p.

Aguon, CF, RE Beck, and MW Ritter. 1992. A method for protecting nests of the Mariana Crow from brown tree snake predation. Proceedings ot the US-Japan Joint Congress on Snake Control for Human Health and Wildlife Conservation. The Snake 24:106.

Aguon, CF. 1996. Presentation to the NRC Committee on the Scientific Bases for Preservation of the Mariana Crow. Guam Department of Agriculture Division Aquatic and Wildlife Resources. May 23, 1996. University of Guam Library.

Aguon, CF, RE Beck Jr, and MW Ritter. In press. A method for protecting nests of the Aga from brown tree snake predation. In GH Rodda, Y Sawai, D Chiszar, and H Tanaka, editors. Snakes, biodiversity, and human health: Case studies in the management of problem snakes.

Anderson, R. 1996. Presentation to the NRC Committee on the Scientific Bases for Preservation of the Mariana Crow. Guam Department of Agriculture Division Aquatic and Wildlife Resources. April 8, 1996. Bishop Museum, Honolulu, HI.

Armstrong, DP and GH Pyke. 1991. Seasonal patterns of reproduction in heathland honeyeaters are not responses to nectar availability. Auk 108:99–107.

Askins, RA and DN Ewert. 1991. Impact of hurricane Hugo on bird populations on St Johns, US Virgin Islands. Biotropica 23:481–7.

Atkinson, C. 1993. Epizootiology of malaria and pox in Hawaiian Crow. Unpubl. Proposal.

Atkinson, IAE. 1985. The spread of commensal species of Rattus to oceanic islands and their effect on island avifaunas. In: PJ Moors, editor. Conservation of island birds. ICBP Tech Publ 3. Washington: Smithsonian Institution Press. p 31–85.

Baker, RH. 1946. Some effects of the war on the wildlife of Micronesia. Trans N Am Wildl Conf 11:205–13.

Baker, RH. 1948. Report on the collections of birds made by United States Naval Medical Research Unit No. 2 in the Pacific war area. Smithsonian Misc Coll 107:1–74.

Baker, RH. 1951. The avifauna of Micronesia, its origin, evolution, and distribution. Lawrence: University of Kansas Publications. 359 p.
Banko, PC. 1996. Telephone conversation on nesting with John Marzluff.
Banko, WE and PC Banko. 1980. History of endemic Hawaiian birds. Part 1. Population histories-species accounts; forest birds; 'Alala or Hawaiian Raven/Crow. Avian History Report no. 6B. Honolulu: University of Hawaii; Cooperative National Park Resource Studies Unit.
Beaty, JJ. 1967. Guam's remarkable birds. S Pacific Bull 17:37-40.
Beck, RE and K Brock. 1996. Presentation to the NRC Committee on the Scientific Bases for Preservation of the Mariana Crow. Guam Department of Agriculture Division Aquatic and Wildlife Resources. May 23, 1996. University of Guam Library.
Beck, RE and CF Aguon. 1996. Presentation to the NRC Committee on the Scientific Bases for Preservation of the Mariana Crow. Guam Department of Agriculture Division Aquatic and Wildlife Resources. May 23, 1996. University of Guam Library.
Bell, HL. 1966. A population study of heathland birds. Emu 65(4):295-304.
Bellingham, PJ, EVJ Tanner, and JR Healy. 1995. Damage and responsiveness of Jamaican montane tree species after disturbance by a hurricane. Ecology 76:2562-80.
Boyce, MS. 1992. Population viability analysis. Ann Rev Ecol Syst 23:481-506.
Brock, K. 1996. Presentation to the NRC Committee on the Scientific Bases for Preservation of the Mariana Crow. Guam Department of Agriculture Division Aquatic and Wildlife Resources. May 23, 1996. University of Guam Library.
Burgman, MA, S Ferson, and HR Akcakaya. 1993. Risk assessment in conservation biology. Chapman and Hall: New York.
[BTSCC] Brown Tree Snake Control Committee. 1996. Brown Tree Snake Control Plan. Report of the Aquatic Nuisance Species Task Force, Brown Tree Snake Control Committee. Honolulu (HI): Fish and Wildlife Service.
Butler, BM. 1991. An archaeological survey of Agiguan (Aguijan) Northern Mariana Islands. Micronesian Archaeological Survey Report 29. Saipan: Division of Historic Preservation. 260 p.
Campbell, EW III. 1996. The effect of brown tree snake (Boiga irregularis) predation on the island of Guam's extant lizard assemblages [PhD dissertation]. Columbus: Ohio State University.
Campbell, EW III. 1996b. Presentation to the NRC Committee on the Scientific Bases for Preservation of the Mariana Crow. National Biological Service. May 23, 1996. Offices of the National Biological Service. Guam National Wildlife Refuge, Guam.
Campbell, EW III, GH Rodda, TH Fritts, and RL Bruggers. Under review. The role of integrated pest management in the control of the brown tree snake (Boiga irregularis) on Pacific Islands. In: GH Rodda, Y Sawai, D Chiszar, and H Tanaka, editors. Snakes, biodiversity, and human health: case studies in the management of problem snakes.
Carano, P and PC Sanchez. 1964. A complete history of Guam. Rutland (VT): Charles E. Tuttle, Inc.
Caswell, H. 1989. Matrix population models. Sunderland (MA): Sinauer Assoc.
Caughley, G. 1994. Directions in conservation biology. J Animal Ecol 63:215-244.
Caughley, G, and A Gunn. 1995. Conservation biology in theory and practice. Cambridge (MA): Blackwell Science, Inc.
Cézilly, F and RG Nager. 1996. Age and breeding performance in monogamous birds: The influence of pair stability. Trends Ecol Evol 11:27.
Chiszar, D. 1990. The behavior of the brown tree snake, Boiga irregularis: A study in applied comparative psychology. In: D. Dewsbury, editor. Contemporary issues in comparative psychology. Sunderland (MA): Sinauer Assoc; p 101-23.
Chiszar, D, K Kandler, R Lee, and HM Smith. 1988. Stimulus control of predatory attack in the brown tree snake (Boiga irregularis). 2. Use of chemical cues during foraging. Amphibia-Reptilia 9:77-88.

Clum, NJ. 1995. Effects of aging and mate retention on reproductive success of captive female Peregrine Falcons. American Zoologist 35:329–339.
Clutton-Brock, TH, editor. 1988. Reproductive success. Chicago: University of Chicago Press.
Collar, NJ, LP Gonzaga, N Krabbe, A Madrono Nieto, LG Naranjo, TA Parker, and DC Wege. 1992. Threatened birds of the Americas. Washington: Smithsonian Institution Press.
Conry, PJ. 1988. High nest predation by brown tree snakes on Guam. Condor 90:478–82.
Craib, JL. 1983. Micronesian prehistory: An archeological overview. Science 219:922–7.
Craig, RJ and E Taisacan. 1994. Notes on the ecology and population decline of the Bridled White-eye. Wilson Bull 106:165–169.
[DAWR] Guam Department of Agriculture Division of Aquatic and Wildlife Resources. 1991. Survey and inventory of non-game birds. Progress Report—Federal Aid to Fish and Wildlife, Proj. No. FW-2R-29, Subproject W-4, Study No. 1, Job. No. 1. 17 p.
[DAWR] Guam Department of Agriculture Division of Aquatic and Wildlife Resources. 1993. Survey and inventory of non-game birds. Progress Report—Federal Aid to Fish and Wildlife, Proj. No. FW-2R-30, Subproject W-4, Study No. 1, Job. No. 1. 17 p.
Delvaux, J. 1991. Reintroduction and recent status of the Raven (Corvus corax L.) in Belgium. In: D Glandt, editor. Der Kolkrabe (Corvus corax) in Mitteleuropa. Metelen, (West Germany): Biologisches Institut Metelen; p 7–111.
Derrickson, SR. 1994. Reintroduction as a component of the Hawaiian Crow recovery program. AZA Ann Conf Proc:82–8.
[DFW] Commonwealth of the Northern Mariana Islands Department of Land and Natural Resources Division of Fish and Wildlife. 1990. Mariana crow ecology. Progress Report—Federal Aid in Wildlife Restoration Program, Project No. W-1-R-8, Study No. IV, Job No. 6. 3 p.
[DFW] Commonwealth of the Northern Mariana Islands Department of Land and Natural Resources Division of Fish and Wildlife. 1993. Mariana crow research. Progress Report—Federal Aid in Wildlife Restoration Program, Project No. W-1-R-1-11, Job No. 6. 10 p.
Diamond, JM and E Mayr. 1976. Species-area relation for the birds of the Soloman archipelago. Proc Natl Acad Sci 73:262–266.
Ebenhard, T. 1988. Introduced mammals and birds and their ecological effects. Swedish Wildl Res 13:1–107.
Eldredge, LG. 1983. Summary of the environmental and fishing information of Guam and the Commonwealth of the Northern Mariana Islands. Historical background, description of the islands, and review of the climate, oceanography, and submarine topography. NOAA Technical Memorandum NMFS 40.
Engbring, J and FL Ramsey. 1984. Distribution and abundance of the forest birds of Guam: Results of a 1981 survey. Washington: US Fish and Wildlife Service; Contract no. FWS-OBS-84/20. 54 p.
Engbring, J and HD Pratt. 1985. Endangered birds in Micronesia: Their history, status, and future prospects. Bird Conservation 2:71–105.
Engbring, J and TH Fritts. 1988. Demise of an insular avifauna: the brown tree snake on Guam. Trans Western Sect Wildl Soc 24:31–7.
Engbring, J, FL Ramsey, and VJ Wildman. 1986. Micronesian forest bird survey, 1982: Saipan, Tinian, Agiguan, and Rota. Honolulu (HI): US Fish and Wildlife Service Special Publication. 143 p.
Fizpatrick, J. 1996. Email communication to Scott Derrickson, May 1996. Cornell Ornithological Laboratory, Ithaca, NY.
Forsman, ED. 1983. Methods and materials for locating and studying spotted owls. Portland (OR): USDA Forest Service General Technical Report no. PNW-162.
Fosberg, FR. 1960. The vegetation of Micronesia. Bull Am Mus Nat Hist 119:1. 75 p.
Fowler, GS. 1995. Stages of age-related and reproductive success in birds: simultaneous effects of age, pair-bond, and reproductive experience. Amer Zool 35:318–328.
Fritts, TH. 1987. Movements of snakes via cargo in the Pacific region. 'Elepaio 47:17–8.

Fritts, TH. 1988. The brown tree snake, Boiga irregularis, a threat to Pacific Islands. Washington: US Fish and Wildlife Service. Biol Rep 88(31).

Fritts, TH and D Chiszar. In press. Snakes on electrical transmission lines: patterns, causes, and strategies for reducing electrical outages due to snakes. In: GH Rodda, Y Sawai, D Chiszar, and H Tanaka, editors. Snakes, biodiversity, and human health: Case studies in the management of problem snakes.

Fritts, TH, MJ McCoid, and DM Gomez. In press. Dispersal of snakes to extralimital islands: incidents of the brown tree snake, Boiga irregularis, dispersing to islands in ships and aircraft. In: GH Rodda, Y Sawai, D Chiszar, and H Tanaka, editors. Snakes, biodiversity, and human health: Case studies in the management of problem snakes.

Fritts, TH, MJ McCoid, and RL Haddock. 1994. Symptoms and circumstances associated with bites by the brown tree snake (Colubridae: Boiga irregularis). Guam. J Herpetol 28:27–33.

Fritts, TH, NJ Scott Jr, and JA Savidge. 1987. Activity of the arboreal brown tree snake (Boiga irregularis) on Guam as determined by electrical power outages. The Snake 19:51–8.

[FWS] US Department of the Interior Fish and Wildlife Service. 1981. Endangered and threatened wildlife and plants: petition acceptance and status review. Federal Register 46(91):26464–9.

[FWS] US Department of the Interior Fish and Wildlife Service. 1990. Native forest birds of Guam and Rota of the Commonwealth of the Northern Mariana Islands Recovery Plan. Portland (OR). 84 p.

[FWS] US Department of the Interior Fish and Wildlife Service. 1991. Endangered and Threatened Wildlife and Plants; Designation of critical habitat for the little Mariana fruit bat, Mariana fruit bat, Guam Broadbill, Mariana Crow, Guam Micronesian Kingfisher, and Guam Bridled White-eye. Federal Register 59(64): 15696–15700.

[FWS] US Department of the Interior Fish and Wildlife Service. 1966. Pacific Islands ecoregion coastal ecosystems program proposal. Honolulu (HI): US Fish and Wildlife Service. 162 p.

[FWS] US Department of the Interior Fish and Wildlife Service. 1984. Endangered and threatened wildlife plants; Determination of Endangered Status for Seven Birds and Two Bats of Guam and the Northern Mariana Islands. Federal Register 49(167): 33881–33885.

[FWS] US Department of the Interior Fish and Wildlife Service. 1994. Endangered and threatened wildlife and plants; withdrawal of proposed rule to designate critical habitat for the Little Mariana Fruit Bat, Mariana Fruit Bat, Guam Broadbill, Aga, Guam Micronesian Kingfisher, and Guam Bridled White-eye. Federal Register 59(64): 15696–700.

Giffin, JG, JM Scott, and S Mountainspring. 1987. Habitat selection and management of the Hawaiian Crow. J Wildl Manag 51:485–94.

Goodwin, D. 1986. Crows of the world. Seattle: University of Washington Press. 299 p.

Greene, HW. 1989. Ecological, evolutionary, and conservation implications of feeding biology in Old World Cat Snakes, genus Boiga (Colubridae). Proc Calif Acad Sci 46:193–207.

Griffith, B, JM Scott, JW Carpenter, and J Reed. 1989. Translocation as a species conservation tool: Status and strategy. Science 245:477–80.

Grout, DJ. 1993. An investigation into the effects of VRC-50 squadron overflights on federally endangered Mariana crows and Mariana fruit bats in Pati Point area of Andersen Air Force Base, Guam. Unpublished report. Honolulu (HI): US Fish and Wildlife Service. 27 p.

Grout, DJ. 1996. Presentation to the NRC Committee on the Scientific Bases for Preservation of the Mariana Crow. US Fish and Wildlife Service. May 21, 1996. Rota, Commonwealth of the Northern Marianas.

Grout, DJ, M Lusk and SG Fancey. 1996. Results of the 1995 Mariana crow survey on Rota. Final report to FWS.

Grue, CE. 1985. Pesticides and the decline of Guam's native birds. Nature 316:301.

Hezel, FX. 1982. From conquest to colonization: The early Spanish mission in the Marianas. J. Pacific Hist 17:115–37.

Hezel, FX and ML Berg, editors. 1981. Micronesia, winds of change. Trust Territory of the Pacific Islands, Saipan. 538 p.

Holmes, DJ, and SN Austad. 1995. The evolution of avian senescence patterns: implications for understanding primary aging processes. Amer Zool 35:307–317.

Jenkins, CD, SA Temple, C van Riper, and WR Hansen. 1989. Disease-related aspects of conserving the endangered Hawaiian Crow. ICBP Tech Publ 10:77–87.

Jenkins, JM. 1979. Natural history of the Guam rail. Condor 81:404–8.

Jenkins, JM. 1983. The native forest birds of Guam. Washington: American Ornithologists' Union. Ornithol Monogr 31.

Kardong, KV, and PR Smith. 1991. The role of sensory receptors in the predatory behavior of the brown tree snake, Boiga irregularis (Squamata: Colubridae). J Herpetol 25:229–31.

Kennedy, PL and DW Stahlecker. 1993. Response of northern goshawks to taped conspecific and Great-Horned Owl calls. J. Raptor Res 24:107–12.

Kleiman, DG, MR Stanley Price, and BB Beck. 1994. Criteria for reintroductions. In: PJS Olney, GM Mace, and ATC Feistner, editors. Creative conservation. London: Chapman & Hall. p 287–303.

Kuehler, C, P Harrity, A Lieberman, and M Kuhn. 1995. Reintroduction of hand-reared alala Corvus hawaiiensis in Hawaii. Oryx 29:261–6.

Kuhlman, M. 1996. Presentation to W. Donald Duckworth and Tania Williams, May 20, 1996. Guam Department of Agriculture, Mangilao, Guam.

Lande, R. 1988. Genetics and demography in biological conservation. Science 241:1455–1460.

Lefkovitch, LP. 1945. The study of population growth in organisms grouped by stages. Biometrics 21:1–18.

Leslie, PH. 1965. On the use of matrices in certain population mathematics. Biometrika 33:183–212.

Lusk, MR and E Taisacan. 1996. Nest measurements for the Mariana crow on Rota, Commonwealth of the Northern Mariana Islands. Unpublished manuscript. Funding provided by Federal Aid in Wildlife Restoration to the CNMI Division of Fisheries and Wildlife; Project no. W-1-R-10. 10 p.

Lynch, JF. 1991. Effects of hurricane Gilbert on birds in a dry tropical forest in the Yucatan Peninsula. Biotropica 23:488–96.

Maben, AF. 1982. The feeding ecology of the black drongo (Dicrurus macrocerus) on Guam. [MS thesis]. Long Beach: California State University.

MacArthur, RH and EO Wilson. 1967. The theory of island biogeography. Princeton (NJ): Princeton University Press.

Marshall, JT Jr. 1949. The endemic avifauna of Saipan, Tinian, Guam and Palau. Condor 51:200–21.

Martin, K. 1995. Patterns and mechanisms for age-dependent reproduction and survival. Amer Zool 35:340–348.

Marzluff, JM. 1993. Use of American Crows and Common Ravens as surrogate species for the endangered Hawaiian Crow. 'Alala Restoration Project 1993 Report. Boise (ID): Peregrine Fund. p 107–48.

Marzluff, JM, KD Whitmore, L Valutis, and L Thomason. 1994. Captive propagation and reintroduction of social birds. Boise (ID): Conservation Research Foundation. 58 p.

Mayr, E. 1945. Bird conservation problems in the Southwest Pacific. Audubon Mag 47:279–82.

McCoid, MJ, TH Fritts, and EW Campbell III. 1994. A brown tree snake (Colubridae: Boiga irregularis) sighting in Texas. Texas J Sci 46:365–8.

McDonald, DB, and H Caswell. 1993. Matrix methods for avian demography. Current Ornithol 10:139–185.

Michael, GA. 1987. Notes on the breeding biology and ecology of the Mariana or Guam crow. Avicult Mag 93:73–82.

Morton, J. 1996a. The effects of aircraft overflights on endangered Mariana crows and Mariana fruit bats at Andersen Air Force Base, Guam. Pearl Harbor: Department of the Navy, Pacific Division, Naval Facilities Engineering Command. 81 p.

Morton, J. 1996b. Presentation to NRC Committee on Scientific Bases for Preservation of the Mariana Crow. US Fish and Wildlife Service. May 23, 1996. Rota, Commonwealth of the Northern Marianas.

Mueller-Dombois, D. 1981. Fire in tropical ecosystems. Fire Regimes and Ecosystem Properties. USDA Forest Service Tech Rep WO-26. p 137–176.

[NOAA] National Oceanic and Atmospheric Administration. 1982. Local climatological data, annual summary with comparative data, Guam, Pacific. Asheville (NC) National Climatic Data Center.

Noon, BR, and JR Sauer. 1992. Population models for passerine birds: structure, parameterization, and analysis. In: DR McCullough and RH Barrett, editors. Wildlife 2001: populations. London: Elsevier Applied Science. p 441–64.

[NRC] National Research Council. 1991. Animals as Sentinels of Environmental Health. Washington: National Academy Press. 160 p.

[NRC] National Research Council. 1992. Scientific Bases for Preservation of the Hawaiian Crow. Washington: National Academy Press. 136 p.

[NRC] National Research Council. 1995. Science and the Endangered Species Act. Washington: National Academy Press. 272 p.

Olsen, G. 1996. Email communication to Bruce Rideout on post-mortem examinations. May 1996. National Biological Service, Patuxent.

Palacios, A. 1996. Presentation to NRC Committee on Scientific Bases for Preservation of the Mariana Crow. CNMI Department of Lands and Natural Resources Division of Fish and Wildlife. April 9, 1996. Bishop Museum, Honolulu, HI.

Pierson, ED, T Elmqvist, WE Rainey, and PA Cox. 1996. Effects of cyclonic storms on flying fox populations on the south Pacific islands of Samoa. Conserv Biol 10:438–51.

Pratt, HD, PL Bruner, and DG Berret. 1979. America's unknown avifauna: The birds of the Mariana Islands. Amer Birds 33:227–35.

Pratt, HD, PL Bruner, and DG Berret. 1987. A field guide to the birds of Hawaii and the tropical Pacific. Princeton (NJ): Princeton University Press.

Ralls, K, K Bruger, and J Ballou. 1980. Inbreeding and juvenile mortality in small populations of ungulates. Science 206:1101–3.

Ralls, K, PH Harvey, and AM Lyles. 1986. Inbreeding in natural populations of birds and mammals. In: ME Soule, editor. Conservation biology: The science scarcity and diversity. Sunderland (MA): Sinauer Assoc. p 35–56.

Ralph, CJ, and HF Sakai. 1979. Forest bird and fruit bat populations and the conservation in Micronesia: notes on a survey. Elepaio 40:20–6.

Ralph, CJ, GR Guepel, P Pyle, TE Martin, and DF DeSante et al. 1993. Handbook of field methods for monitoring land birds. Albany (CA): USDA Forest Service. General Technical Report nr PSW-GTR-144. 41 p.

Reese, KP, and JA Kadlec. 1985. Influence of high density and parental age on the habitat selection and reproduction of black-billed magpies. Condor 87:96–105.

Reichel, JD and PO Glass. 1991. Checklist of the birds of the Mariana Islands. Elepaio 51:3–11.

Reichel, JD, GJ Wiles, and PO Glass. 1992. Island extinctions: the case of the endangered nightingale reed warbler. Wilson Bull 104:44–54.

Renssen, TA. 1988. Herintroductie van de Raff Corvus corax in Nederland. Limosa 61:137–44.

Reynolds, RT, JM Scott, and RA Nussbaum. 1980. A variable circular-plot method for estimating bird numbers. Condor 82:309–313.

Ramsey, FL and JM Scott. 1981. Analysis of bird survey data using a modification of Emlen's methods. In: CJ Ralph and JM Scott, editors. Estimating numbers of terrestrial birds. Studies Avian Biol 6:483–7.

Ricklefs, RE. 1969. An analysis of nesting mortality in birds. Smithsonian Contrib. Zool. 9:1–48.

Rodda, GH. 1996. Presentation to NRC Committee on Scientific Bases for Preservation of the Mariana Crow. National Biological Service. May 23, 1996. National Biological Service, Guam National Wildlife Refuge, Guam.

Rodda, GH and TH Fritts. 1992. The impact of the introduction of the brown tree snake, *Boiga irregularis*, on Guam's lizards. J Herpetol 26:166–74.

Rodda, GH and TH Fritts. 1993. The brown tree snake on Pacific Islands: 1993 status. Pac Sci Asso Info Bull 45(3–4):1–3.

Rodda, GH, MJ McCoid, TH Fritts, and EW Campbell III. In press (c). Population trends and limiting factors in *Boiga irregularis*. In: GH Rodda, Y Sawai, D Chiszar, and H Tanaka, editors. Snakes, biodiversity, and human health: Case studies in the management of problem snakes.

Rodda, GH, TH Fritts, and EW Campbell III. In press (b). The feasibility of controlling the brown tree snake in small plots. In: GH Rodda, Y Sawai, D Chiszar, and H Tanaka, editors. Snakes, biodiversity, and human health: Case studies in the management of problem snakes.

Rodda, GH, TH Fritts, and PJ Conry. 1992. Origin and population growth of the brown tree snake, *Boiga irregularis*, on Guam. Pacific Sci 24:46–57.

Rodda, GH, TH Fritts, MJ Mccoid, and EW Campbell III. In press (a). An overview of the biology of the brown tree snake, Boiga irregularis, a costly introduced pest on Pacific Islands. In: GH Rodda, Y Sawai, D Chiszar, and H Tanaka, editors. Snakes, biodiversity, and human health: Case studies in the management of problem snakes.

Romanoff, AL and AJ Romanoff. 1972. Pathogenesis of the avian embryo. J. New York: Wiley and Sons.

Røskaft, E, Y Espmark, and T Jarvi. 1983. Reproductive effort and breeding success in relation to age by the Rook *Corvus fugilegus*. Ornis Scand 14:169–74.

Safford, WE. 1902. Birds of the Mariana Islands and their vernacular names. Osprey 6:39–42, 65–70.

Sakai, HF and JR Carpenter. 1990. The variety and nutritional value of foods consumed by Hawaiian Crow nestlings, an endangered species. Condor 92:220–8.

Sakai, HF, CJ Ralph, and CD Jenkins. 1986. Foraging ecology of the Hawaiian Crow, and endangered generalist. Condor 88:211–9.

Sæther, BK. 1990. Age-specific variation in reproductive performance of birds. Current Ornithol 7:251–83.

Savidge, JA. 1984. Guam: paradise lost for wildlife. Biol Conserv 30:305–17.

Savidge, JA. 1986. The role of disease and predation in the decline of Guam's avifauna [dissertation]. Urbana-Champaign: University of Illinois. 79 p.

Savidge, JA. 1987. Extinction of an island avifauna by an introduced snake. Ecology 68:660–8.

Savidge, JA. 1988. Food habits of *Boiga irregularis*, an introduced predator on Guam. J Herpetol 22:275–82.

Scatena, FN and MC Larsen. 1991. Physical aspects of Hurricane Hugo in Puerto Rico. Biotropica 23:317–23.

Schoenwald-Cox, CM, SM Chambers, B MacBryde, and WL Thomas. 1983. Genetics and conservation: A reference for managing wild animal and plant populations. Menlo Park (CA): Benjamin/Cummings.

Seale, A. 1901. Report of a mission to Guam. Honolulu (HI): Occassional Papers of the Bernice P Bishop Museum 1:17–128.

Shields, WM. 1993. The natural and unnatural history of inbreeding and outbreeding. In Thornhill, NW, editor. The natural history of inbreeding and outbreeding; theoretical and empirical perspectives. Chicago: University of Chicago Press.

Shine, R. 1991a. Strangers in a strange land: ecology of Australian colubrid snakes. Copeia 1991:120–31.

Shine, R. 1991b. Australian snakes: A natural history. Ithaca (NY): Cornell University Press.

Snyder, NFR, SR Derrickson, SR Beissinger, JW Wiley, TB Smith, WD Toone, and B Miller. 1996. Limitations of captive breeding in endangered species biology. Cons Biol 10:338–48.

Solenberger, RR. 1967. The changing role of rice in the Northern Mariana Islands. Micronesica 3:97–103.

Soulé, ME, editor. 1987. Viable populations for conservation. New York: Cambridge University Press.

Steadman, DW. 1992. Extinct and extirpated birds from Rota, Mariana Islands. Micronesica 25:71–84.

Steadman, DW. 1995a. Prehistoric extinction of Pacific island birds. Science 267:1123–31.

Steadman, DW. 1995b. Determining the natural distribution of resident birds in the Mariana Islands. [Unpublished]. Preliminary Report to Fish and Wildlife Service. Honolulu (HI): US Fish and Wildlife Service. 57 p.

Stoleson, SH. 1996. Hatching asynchrony in the Green-rumped parrotlet: A multiple hypothesis analysis [PhD dissertation]. New Haven (CT): Yale University.

Stoleson, SH and SR Beissinger. 1995. Hatching asynchrony and the onset of incubation in birds. Current Ornithol 12:191–271.

Stone, BC. 1970. The flora of Guam. Micronesica 6:1–659.

Strophlet, JJ. 1946. Birds of Guam. Auk 53:539–40.

Tarr, CL and RC Fleischer. 1995. A molecular assessment of genetic variability and population differentiation in the endangered Mariana crow (Corvus kubaryi). Honolulu (HI): US Fish and Wildlife Service. Available from the National Zoological Park, Smithsonian, Washington.

Tomback, DF. 1986. Observations on the behavior and ecology of the Mariana crow. Condor 88:398–401.

Unger, K. 1996. Presentation to the NRC Committee on Scientific Bases for Preservation of the Mariana Crow. McCandless Ranch, Waimae, Big Island, HI.

vanRiper, C, SG van Riper, ML Goff, and M Laird. 1986. The epizootiology and ecological significance of malaria in Hawaiian land birds. Ecol Monogr 56:327–44.

Veiga, JP. 1992. Hatching asynchrony in the House Sparrow: a test of the egg-viability hypothesis. Amer Natural. 139:669–75.

Waide, RB. 1991. The effect of hurricane Hugo on bird populations in the Luquillo experimental forest, Puerto Rico. Biotropica 23:475–80.

Walker. LR. 1991. Tree damage and recovery from hurricane Hugo in Luquillo experimental forest, Puerto Rico. Biotropica 23:379–85.

Wauer, RH and JM Wunderle Jr. 1992. The effect of hurricane Hugo on bird populations on St. Croix, US Virgin Islands. Wilson Bull 104:656–73.

Webb, DR. 1987. Thermal tolerance of avian embryos: a review. Condor 89:874–98.

White, FN and JL Kinney. 1974. Avian incubation. Science 186:107–15.

Wiles, GJ. 1987. Current research and future management of Marianas Fruit Bats (Chiroptera: *Pteropodidae*) on Guam. Aust Mammal 10:93–5.

Wiles, GJ, CF Aguon, MK Brock, and RE Beck. 1994. Funding needs for the conservation of the Mariana crow (*Corvus kubaryi*) on Guam for 1995–1996. Division of Aquatic and Wildlife Resources, Unpublished Report to USFWS, August 1994. 26 p.

Wiles, GL, CF Aguon, GW Davis, and DJ Grout. 1995. The status and distribution of endangered animals and plants in northern Guam. Micronesica 28:31–49.

Wiles, GL, GH Rodda, TH Fritts, and S Taisacan. 1990. Abundance and habitat use of reptiles on Rota, Mariana Islands. Micronesica 23:153–66.

Wiley, TR and JM Wunderle Jr. 1994. The effects of hurricanes on birds, with special reference to Caribbean islands. Bird Cons Intern 4:1–31.

Will, T. 1991. Birds of a severely hurricane-damaged Atlantic coast rain forest in Nicaragua. Biotropica 23:497–507.

Woolfenden, GE and JW Fitzpatrick. 1984. The Florida Scrub Jay. Princeton (NJ): Princeton University Press. 406 p.

Worthington, DJ and EM Taisacan. 1995. Mariana crow research, fiscal year 1995. Federal Aid in Wildlife Restoration Program. Commonwealth of the Northern Mariana Islands. Division of Fish and Wildlife, Project W-1-R-1-11, Job 6. 6 p.

Wunderle, JM Jr. 1995. Responses of bird populations in a Puerto Rican Forest to hurricane Hugo: The first 18 months. Condor 97:879–96.

Wunderle, JM Jr., DJ Lodge, and RB Waide. 1992. Short-term effects of hurricane Gilbert on terrestrial bird populations on Jamaica. Auk 109:148–66.

Appendixes

Appendix A

Statement of Task

A committee of about eight experts in ornithology, population biology, and captive propagation, including, where appropriate, members of the previous NRC Committee on Scientific Bases for the Preservation of the Hawaiian Crow, will analyze the existing scientific information on and alternative approaches to the recovery of the Mariana crow and will prepare a report detailing their findings, conclusions, and recommendations concerning the recovery efforts for the species. The committee will examine existing scientific data, hold a public meeting, and speak to individuals involved in the recovery effort. The committee will meet three times. One meeting in Guam and Rota will be planned to review data, visit field sites, and gather other relevant information for the committee's deliberations. The committee will address scientific issues such as the following:

- Causes of the continuing decline in population of the Mariana crow in the wild;
- Options for action to halt or reverse the decrease in numbers of the Mariana crow;
- Approximate minimum viable population for survival of the Guam population of this species; and
- The advisability of adding genetic material from the Rota population to the Guam population.

The committee will then develop scientific recommendations that will assist the interested parties in working towards the recovery of *Corvus kubaryi*.

Appendix B

Acknowledgments

The committee acknowledges with appreciation presentations made at meetings of the community and personal communications by the following persons:

Guam Department of Agriculture
Mike Kuhlman, Director
Celestino Aguon, Wildlife Biologist, Division of Aquatic and Wildlife Resources
Robert Anderson, Chief, Division of Aquatic and Wildlife Resources
Robert Beck, Wildlife Supervisor, Division of Aquatic and Wildlife Resources
Kelly Brock, Wildlife Biologist, Division of Aquatic and Wildlife Resources
Gary Wiles, Wildlife Biologist, Division of Aquatic and Wildlife Resources

Commonwealth of the Northern Mariana's Department of Lands and Natural Resources
Benigno Sablan, Secretary
Arnold Palacius, Director, Division of Fisheries and Wildlife
Frank Toves, Resident Director, Rota
Stan Taisacan, Wildlife Biologist, Division of Fisheries and Wildlife

U.S. Fish and Wildlife Service
Dan Grout, Wildlife Biologist, Rota
John Engbring, Wildlife Biologist, Olympia WA
Brooks Harper, Division Chief, Honolulu HI
Michael Lusk, Wildlife Biologist, Honolulu HI
John Morton, Wildlife Biologist, Guam
Karen Rosa, Chief, Branch of Recovery, Honolulu HI
Robert Smith, Manager, Pacific Islands Ecoregion, Honolulu HI

And the many other individuals and organizations who provided assistance
Anderson Air Force Base
Kristy Anderson, National Biological Service, Guam
Bishop Museum staff
Issac Calvo, Chair, Rota Habitat Conservation Committee
Stephen Camacho, Wildlife Technician, Rota Mayor's Office
Earl Campbell, National Biological Service, Columbus OH
Rick Engemann, USDA Animal Damage Control, Ft. Collins CO
John Fitzpatrick, Cornell University
Robert Fleischer, National Zoological Park, Washington DC
Blu Gamaio, Bishop Museum, Honolulu HI
David Gillespie, Dean, University of Guam
Vince Leon Guerro, US Rep. Underwood's office, Guam
Peter Harrity, The Peregrine Fund, Volcano HI
Heidi Hirsh, Andersen Air Force Base, Guam
Anne Marshall, Honolulu, HI
Aric Noboa, US Rep. Underwood's office, Washington DC
Gordon Rodda, National Biological Service, Guam
Cynthia Salley, McCandless Ranch, HI
Cheryl Tarr, National Zoological Park, Washington DC
The Honorable Robert P. Underwood, Guam Delegate to US Congress
Keith Unger, McCandless Ranch, HI
David Worthington, Wildlife Biologist, Rota

Appendix C

Committee and Staff Biographical Information

W. Donald Duckworth, Chair, is president and director of the Bishop Museum in Honolulu, Hawaii. Trained in systematic entomology, his research has focused on lepidoptera in the tropics.

Steven R. Beissinger is an associate professor of conservation biology at the University of California, Berkeley. His research has focused on conservation and ecology of endangered or exploited species, demographic models of population viability and recovery, and assessing ecosystem conservation priorities, as well as other areas.

Scott R. Derrickson is curator of ornithology and deputy associate director for conservation, Smithsonian Institution's National Zoological Park. His research has focused primarily on endangered-species propagation and reintroduction.

Thomas H. Fritts is a wildlife biologist and chief of the Biological Survey Program, Patuxent Wildlife Research Center, in the National Biological Service, and is curator of amphibians and reptiles in the Department of Vertebrate Zoology of the Smithsonian Institution's National Museum of Natural History. His research has focused on Pacific Island ecology with special emphasis on the biology of the brown tree snake, problems it causes, and potential strategies for its control.

Susan M. Haig is the senior wildlife ecologist at the Forest and Rangeland Ecosystem Science Center (National Biological Service) and an associate professor at Oregon State University. Her research focuses on genetics and demography of small populations and shorebird behavioral ecology.

Frances C. James is a professor in the Department of Biological Science at Florida State University. Her research has investigated biogeographic questions in ornithology, including habitat relationships, intraspecific geographic variation in morphology, and long-term trends in populations.

John M. Marzluff is the senior scientist at Sustainable Ecosystems Institute in Boise, Idaho. His research interests included the biology of corvids (crows, ravens, and jays), specifically how their increasing populations in North America lead to increases in nest predation rates on threatened species, and how to restore populations of endangered crows on Pacific islands.

Bruce Rideout is a senior pathologist in the Center for Reproduction of Endangered Species of the Zoological Society of San Diego. His primary research interests include avian embryonic and neonatal pathology as it relates to captive propagation for recovery programs, avian infectious diseases, and the risk of disease transmission in translocation and reintroduction programs.

Staff

Tania Williams is a program officer in the Board on Biology, Commission on Life Sciences of the National Research Council. Her interests include natural resource management, biodiversity studies, and science policy.